T0261544

Yamaha 250 and 350 Twins Owners Workshop Manual

by Jeff Clew

Models covered
YDS7. 247cc. UK December 1970 to June 1973
RD250. 247cc. UK and US July 1973 to May 1975
RD250 DX. 247cc. UK January 1976 to 1979
YR5. 347cc. UK December 1970 to July 1973
RD350. 347cc. UK and US July 1973 to May 1975
RD350 B. 347cc. UK May 1975 to September 1976

ISBN 978 0 85696 505 0

© Haynes Publishing 2017

ABCDE
FGHIJ
KLMNO
PQRS

3

Haynes Publishing
Sparkford Nr Yeovil
Somerset BA22 7JJ England

Haynes Publications, Inc
859 Lawrence Drive
Newbury Park
California 91320 USA

Acknowledgements

Our grateful thanks are due to Mr Lawrie Hockley, of W and H Brockliss Limited, 334, Brockley Road, London, SE4 who provided much useful information based on his wide experience as a Yamaha repair specialist. Brian Horsfall gave the necessary assistance with the overhaul and devised the ingenious methods for overcoming the lack of service tools. Les Brazier arranged and took the photographs that accompany the text. Tim Parker edited the text.

Our thanks are due also to Brian Hoppé, of Leliott and Bird, Sherborne, Dorset, who located the YR5 model featured in this manual, and to Bryan Goss Motor Cycles, of Yeovil, who obtained all the necessary spare parts. Finally, we would also like to acknowledge the help of the Avon Rubber Company, who kindly supplied illustrations and advice about tyre fitting and NGK Spark Plug (UK) Limited for the provision of spark plug photographs.

The front cover transparency of the RD250 DX was supplied by Mitsui Machinery Sales (UK) Ltd.

About this manual

The author of this manual has the conviction that the only way in which a meaningful and easy to follow text can be written is first to do the work himself, under conditions similar to those found in the average household. As a result, the hands seen in the photographs are those of the author. Even the machines are not new: examples that have covered a considerable mileage were selected so that the conditions encountered would be typical of those found by the average owner. Unless specially mentioned, and therefore considered essential, Yamaha special service tools have not been used. There is invariably some alternative means of loosening or removing a vital component when service tools are not available but risk of damage should always be avoided.

Each of the six Chapters is divided into numbered sections. Within these sections are numbered paragraphs. Cross reference throughout the manual is quite straightforward and logical. When reference is made "See Section 6.10' it means Section 6, paragraph 10 in the same chapter. If another chapter were meant, the reference would read "See Chapter 2, Section 6.10".

All the photographs are captioned with a section/paragraph number to which they refer, and are relevant to the chapter text adjacent.

Figures (usually line illustrations) appear in a logical but numerical order, within a given chapter. Fig.1.1 therefore refers to the first figure in Chapter 1.

Left-hand and right-hand descriptions of the machines and their components refer to the left and right of a given machine when the rider is seated normally.

Motorcycle manufacturers continually make changes to specifications and recommendations, and these, when notified, are incorporated into our manuals at the earliest opportunity.

We take great pride in the accuracy of information given in this manual, but motorcycle manufacturers make alterations and design changes during the production run of a particular motorcycle of which they do not inform us. No liability can be accepted by the authors or publishers for loss, damage or injury caused by any errors in, or omissions from, the information given.

Modifications to the Yamaha 250/350 cc range

The first 250 cc twin cylinder model to appear in the UK was the YDS3, which became available during November 1964. Several years passed before the matching 350 cc version was added to the UK range; the YR3 model in March 1969. The most significant design changes occurred during December 1970, when the new YDS7 and YR5 models were imported into the UK. These models use a duplex cradle frame of improved design and have the engine/gear unit arranged with a horizontal crankcase split, superceding the earlier vertical arrangement. Refer to Chapter 7 for the RD models.

Dimensions

	YDS7	YR5
Overall length	80.3 in.	80.3 in.
Overall width	32.9 in.	32.9 in.
Overall height	42.7 in.	42.7 in.
Wheelbase	52.0 in.	52.0 in.
Minimum ground clearance	5.9 in.	6.1 in.
Weight (dry)	304 lbs	308 lbs

Contents

The Yamaha 350 cc (1971 model)

1979 RD250DX

Introduction to the Yamaha 250/350 twins

Although the history of Yamaha can be traced back to the year 1887, when a then very small company commenced manufacture of reed organs, it was not until 1954 that the company became interested in motor cycles. As can be imagined, the problems of marketing a motor cycle against a background of musical instruments manufacture were considerable. Some local racing successes and the use of hitherto unknown bright colour schemes helped achieve the desired results and in July 1955 the Yamaha Motor Company was established as a separate entity, employing a work force of less than 100 and turning out some 300 machines a month.

Competition successes continued and with the advent of tasteful styling that followed Italian trends, Yamaha became established as one of the world's leading motor cycle manufacturers. Part of this success story is the impressive list of Yamaha 'firsts' - a whole string of innovations that include electric starting, pressed steel frame, torque induction and 6 and 8 port engines. There is also the "Autolube" system of lubrication, in which the engine-driven oil pump is linked to the twist grip throttle, so that lubrication requirements are always in step with engine demands.

Since 1964, Yamaha has gained the World Championship on numerous occasions, in both the 125 cc and 250 cc classes. Indeed, Yamaha has dominated the lightweight classes in international road racing events to such an extent in recent years that several race promotors are now instituting a special type of event in their programme from which Yamaha machines are barred! Most of the racing successes have been achieved with twin cylinder two-strokes and the practical experience gained has been applied to the road going versions such as the YDS7 and YR5 models and their later variants, the reed valve induction RD models.

Ordering spare parts

When ordering spare parts for the Yamaha 250/350 cc twins, it is advisable to deal direct with an official Yamaha agent, who will be able to supply many of the items required ex-stock. Although parts can be ordered from Yamaha direct, it is preferable to route the order via a local agent even if the parts are not available from stock. He is in a better position to specify exactly the parts required and to identify the relevant spare part numbers so that there is less chance of the wrong part being supplied by the manufacturer due to a vague or incomplete description.

When ordering spares, always quote the frame and engine numbers in full, together with any prefixes or suffixes in the form of letters. The frame number is found stamped on the right-hand side of the steering head, in line with the forks. The engine number is stamped on the left-hand side of the upper crankcase, immediately below the left-hand carburettor.

Use only parts of genuine Yamaha manufacture. A few pattern parts are available, sometimes at cheaper prices, but there is no guarantee that they will give such good service as the originals they replace. Retain any worn or broken parts until the replacements have been obtained; they are sometimes needed as a pattern to help identify the correct replacement when design changes have been made during a production run.

Some of the more expendable parts such as spark plugs, bulbs, tyres, oils and greases etc., can be obtained from accessory shops and motor factors, who have convenient opening hours, charge lower prices and can often be found not far from home. It is also possible to obtain parts on a Mail Order basis from a number of specialists who advertise regularly in the motor cycle magazines.

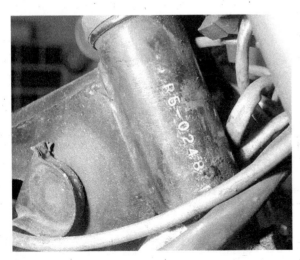

Frame number location

Engine number location

Safety first!

Professional motor mechanics are trained in safe working procedures. However enthusiastic you may be about getting on with the job in hand, do take the time to ensure that your safety is not put at risk. A moment's lack of attention can result in an accident, as can failure to observe certain elementary precautions.

There will always be new ways of having accidents, and the following points do not pretend to be a comprehensive list of all dangers; they are intended rather to make you aware of the risks and to encourage a safety-conscious approach to all work you carry out on your vehicle.

Essential DOs and DON'Ts

DON'T start the engine without first ascertaining that the transmission is in neutral.

DON'T suddenly remove the filler cap from a hot cooling system – cover it with a cloth and release the pressure gradually first, or you may get scalded by escaping coolant.

DON'T attempt to drain oil until you are sure it has cooled sufficiently to avoid scalding you.

DON'T grasp any part of the engine, exhaust or silencer without first ascertaining that it is sufficiently cool to avoid burning you.

DON'T allow brake fluid or antifreeze to contact the machine's paintwork or plastic components.

DON'T syphon toxic liquids such as fuel, brake fluid or antifreeze by mouth, or allow them to remain on your skin.

DON'T inhale dust – it may be injurious to health (see *Asbestos* heading).

DON'T allow any spilt oil or grease to remain on the floor – wipe it up straight away, before someone slips on it.

DON'T use ill-fitting spanners or other tools which may slip and cause injury.

DON'T attempt to lift a heavy component which may be beyond your capability – get assistance.

DON'T rush to finish a job, or take unverified short cuts.

DON'T allow children or animals in or around an unattended vehicle.

DON'T inflate a tyre to a pressure above the recommended maximum. Apart from overstressing the carcase and wheel rim, in extreme cases the tyre may blow off forcibly.

DO ensure that the machine is supported securely at all times. This is especially important when the machine is blocked up to aid wheel or fork removal.

DO take care when attempting to slacken a stubborn nut or bolt. It is generally better to pull on a spanner, rather than push, so that if slippage occurs you fall away from the machine rather than on to it.

DO wear eye protection when using power tools such as drill, sander, bench grinder etc.

DO use a barrier cream on your hands prior to undertaking dirty jobs – it will protect your skin from infection as well as making the dirt easier to remove afterwards; but make sure your hands aren't left slippery. Note that long-term contact with used engine oil can be a health hazard.

DO keep loose clothing (cuffs, tie etc) and long hair well out of the way of moving mechanical parts.

DO remove rings, wristwatch etc, before working on the vehicle – especially the electrical system.

DO keep your work area tidy – it is only too easy to fall over articles left lying around.

DO exercise caution when compressing springs for removal or installation. Ensure that the tension is applied and released in a controlled manner, using suitable tools which preclude the possibility of the spring escaping violently.

DO ensure that any lifting tackle used has a safe working load rating adequate for the job.

DO get someone to check periodically that all is well, when working alone on the vehicle.

DO carry out work in a logical sequence and check that everything is correctly assembled and tightened afterwards.

DO remember that your vehicle's safety affects that of yourself and others. If in doubt on any point, get specialist advice.

IF, in spite of following these precautions, you are unfortunate enough to injure yourself, seek medical attention as soon as possible.

Asbestos

Certain friction, insulating, sealing, and other products – such as brake linings, clutch linings, gaskets, etc – contain asbestos. *Extreme care must be taken to avoid inhalation of dust from such products since it is hazardous to health.* If in doubt, assume that they *do* contain asbestos.

Fire

Remember at all times that petrol (gasoline) is highly flammable. Never smoke, or have any kind of naked flame around, when working on the vehicle. But the risk does not end there – a spark caused by an electrical short-circuit, by two metal surfaces contacting each other, by careless use of tools, or even by static electricity built up in your body under certain conditions, can ignite petrol vapour, which in a confined space is highly explosive.

Always disconnect the battery earth (ground) terminal before working on any part of the fuel or electrical system, and never risk spilling fuel on to a hot engine or exhaust.

It is recommended that a fire extinguisher of a type suitable for fuel and electrical fires is kept handy in the garage or workplace at all times. Never try to extinguish a fuel or electrical fire with water.

Note: *Any reference to a 'torch' appearing in this manual should always be taken to mean a hand-held battery-operated electric lamp or flashlight. It does* **not** *mean a welding/gas torch or blowlamp.*

Fumes

Certain fumes are highly toxic and can quickly cause unconsciousness and even death if inhaled to any extent. Petrol (gasoline) vapour comes into this category, as do the vapours from certain solvents such as trichloroethylene. Any draining or pouring of such volatile fluids should be done in a well ventilated area.

When using cleaning fluids and solvents, read the instructions carefully. Never use materials from unmarked containers – they may give off poisonous vapours.

Never run the engine of a motor vehicle in an enclosed space such as a garage. Exhaust fumes contain carbon monoxide which is extremely poisonous; if you need to run the engine, always do so in the open air or at least have the rear of the vehicle outside the workplace.

The battery

Never cause a spark, or allow a naked light, near the vehicle's battery. It will normally be giving off a certain amount of hydrogen gas, which is highly explosive.

Always disconnect the battery earth (ground) terminal before working on the fuel or electrical systems.

If possible, loosen the filler plugs or cover when charging the battery from an external source. Do not charge at an excessive rate or the battery may burst.

Take care when topping up and when carrying the battery. The acid electrolyte, even when diluted, is very corrosive and should not be allowed to contact the eyes or skin.

If you ever need to prepare electrolyte yourself, always add the acid slowly to the water, and never the other way round. Protect against splashes by wearing rubber gloves and goggles.

Mains electricity and electrical equipment

When using an electric power tool, inspection light etc, always ensure that the appliance is correctly connected to its plug and that, where necessary, it is properly earthed (grounded). Do not use such appliances in damp conditions and, again, beware of creating a spark or applying excessive heat in the vicinity of fuel or fuel vapour. Also ensure that the appliances meet the relevant national safety standards.

Ignition HT voltage

A severe electric shock can result from touching certain parts of the ignition system, such as the HT leads, when the engine is running or being cranked, particularly if components are damp or the insulation is defective. Where an electronic ignition system is fitted, the HT voltage is much higher and could prove fatal.

Routine Maintenance

Periodic routine maintenance is a continuous process that commences immediately the machine is used. It must be carried out at specified mileage recordings, or on a calendar basis if the machine is not used frequently, whichever is the sooner. Maintenance should be regarded as an insurance policy, to help keep the machine in the peak of condition and to ensure long, trouble-free service. It has the additional benefit of giving early warning of any faults that may develop and will act as a regular safety check, to the obvious advantage of both rider and machine alike.

The various maintenance tasks are described under their respective mileage and calendar headings. Accompanying diagrams are provided, where necessary. It should be remembered that the interval between the various maintenance tasks serves only as a guide. As the machine gets older or is used under particularly adverse conditions, it would be advisable to reduce the period between each check. Some of the tasks are described in detail, where they are not mentioned fully as a routine maintenance item in the text. If a specific item is mentioned but not described in detail, it will be covered fully in the appropriate Chapter. No special tools are required for the normal routine maintenance tasks. The tools supplied with every new machine will prove adequate, or if they are not available, the tools found in the average household.

Weekly, or every 200 miles

Check the adjustment of the front and rear brakes.
Check the oil level in the gearbox and top up if necessary.
Check the oil level in the oil tank and top up if necessary.
Oil all exposed control cables and joints.
Check the tyre pressures.
Check the electrolyte level of the battery and top up with distilled water, if necessary. DO NOT OVERFILL.

Three-monthly, or every 2,000 miles

Complete all the checks listed in the weekly/200 mile service and then the following items:-
Apply a grease gun to all the greasing points.
Remove, inspect and clean both spark plugs. If necessary, re-gap the points.
Clean the contact breaker points and re-set the gaps.
Verify the accuracy of the ignition timing by the alternator timing marks.
Decarbonise the cylinder heads and the exhaust ports.
Remove and clean the silencer baffles.
Remove, clean and lubricate the final drive chain.
Clean or renew the air cleaner element.
Check the setting of the "Autolube" oil pump and re-adjust if necessary. This is the dealer's responsibility whilst the machine is still under guarantee. Unauthorised tampering may invalidate the guarantee.

Six monthly, or every 4,000 miles

Complete all the checks listed in the weekly and three-monthly services, then dismantle and clean both carburettors.

Oil changes

Although a two-stroke engine runs on the total loss principle and does not have an oil content which is recirculated and retained in the crankcase, there is need to change the gearbox oil at regular intervals. The recommended period between oil changes is every 1,000 miles or six weeks, whichever is the sooner. If the machine is used for a succession of short journeys, or when tem-

peratures are low, it is advisable to halve this period, to offset the effects of condensation.

Statutory requirements

Although no specific mention has been made of the tyres, lighting equipment, horn or speedometer, statutory requirements relate to the depth of tyre tread permissible before replacement must be made and to the correct functioning of the lighting equipment, horn and speedometer. Whilst the prudent rider will check each of these points during the course of the routine maintenance tasks, as well as the general security of all nuts and bolts etc, it should be remembered that failure to observe the regulations may result in prosecution. Apart from the legal implications, it is best to err on the side of safety.

RM1. Dipstick indicates oil level in tank

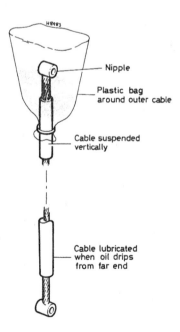

RM2. Control cable oiling

Quick glance: Maintenance adjustments and capacities

Engine	SAE 30 two-stroke oil. Separate lubrication system operated by 'Autolube' oil pump, interconnected with the throttle
Oil tank	Capacity 3½ Imp. pints (2 litres)
Gearbox	Capacity 2.6 Imp. pints (1½ litres) SAE 10W/30 engine oil
Front forks	145 cc per fork leg SAE 10W/30 engine oil
Contact breaker gap	0.3 to 0.4 mm (0.012 in to 0.016 in)
Sparking plug gap	0.5 - 0.7 mm (0.020 - 0.030 in)
Tyre pressures	23 psi front, 29 psi rear For continuous high speed riding or when a pillion passenger is carried, add 5 psi to the above recommendations, for both tyres.

Recommended lubricants

Component	Grade	Castrol Grade
Engine	SAE 30 two-stroke oil	**Castrol TT Two-Stroke Oil**
Gearbox and Forks	SAE 10W/30 engine oil	**Castrolite Multigrade**
Final Drive Chain	Multi-grade oil or graphited grease	**Castrol GTX** **Castrol Graphited Grease**
All Greasing Points	Multi-purpose high melting point lithium	**Castrol LM Grease**

Line drawings

Chapter 1 Engine, clutch and gearbox

Contents

Specifications

Engine

	YSD7	YR5
Type	Twin cylinder, two-stroke, air cooled	
Porting	six port*	
Capacity	247 cc	347 cc
Bore	54 mm	64 mm
Stroke	54 mm	54 mm
Compression ratio	7.1 : 1	6.9 : 1
bhp	30 @ 7,500 rpm	36 @ 7,000 rpm
Lubrication	Separate oil pump using 'Autolube' system	

*Referred to by Yamaha as 5 port; the exhaust port is not counted.

Pistons

Clearance in cylinder bore	0.040 to 0.045 mm (0.0016 in to 0.0018 in)
Wear limit	0.050 mm (0.0020 in)
Rebore sizes available	+0.25 mm and +0.50 mm (0.010 in and 0.020 in)

Piston rings

Number per piston ...	...	...	...	...	...	...		Two
Eng gaps ...	...	...	...	...	...	...		0.45 mm to 0.65 mm
								(0.017 to 0.025 in)

Gearbox

Type	...	...	...	...	...	...	...		Five-speed, constant mesh	
Ratios:										
Bottom gear	...	...	...	...	...	...	...	22.126 : 1		19.608 : 1
2nd gear ...	...	...	...	...	...	...	...	13.737 : 1		12.174 : 1
3rd gear ...	...	...	...	...	...	...	...	10.295 : 1		9.123 : 1
4th gear ...	...	...	...	...	...	...	...	8.337 : 1		7.338 : 1
Top gear ...	...	...	...	...	...	...	...	6.963 : 1		6.171 : 1

Clutch

Type	...	...	...	...	...	...		Wet, multi-plate type
No. of plates:								
Plain ...	...	...	...	...	...	...		Seven
Inserted ...	...	...	...	...	...	...		Six
Clutch springs	...	...	...	...	...	...		Six
Free length	...	...	...	...	...	...		36 mm
Wear limit	...	...	...	...	...	...		35 mm
Inserted clutch plate thickness			...	...	...	...		3.0 mm
Wear limit	...	...	...	...	...	...		2.7 mm

1 General description

The engine unit fitted to the Yamaha 250/350 cc YDS7 and YR5 models is a twin cylinder two-stroke, using flat top pistons and what is known as the loop scavenging system to effect a satisfactory induction and exhaust sequence. Each cylinder barrel has six ports (inlet, exhaust, two transfer and two auxilliary transfer) the design of which permits the incoming charge to expel the exhaust gases and at the same time cool the piston. The piston rings are pegged in characteristic two-stroke practice, to prevent them from rotating and the ends being trapped in the cylinder barrel ports. Large diameter oil seals form an effective seal around the built-up crankshaft assembly, which runs on four journal ball bearings and has full flywheels. All engine/gear castings are in aluminium alloy, including the cylinder barrels and cylinder heads. The crankcase is arranged to separate horizontally, simplifying dismantling and reassembly operations. Because the engine and gearbox are built in unit, when the engine is dismantled the gearbox has to be dismantled too, and vice-versa.

The generator is on the left-hand side of the engine. It is an alternator with the rotor attached to the end of the crankshaft. The clutch assembly is located on the right-hand side, behind the cover containing the kickstarter mechanism. The exhaust system is twin downswept, and each cylinder has its own exhaust pipe and silencer. The gearchange pedal is on the left-hand side of the machine; it operates a five-speed constant mesh gearbox.

Lubrication is effected by the Yamaha "Autolube" system, which takes the form of a gear driven oil pump drawing oil from a separate oil tank and distributing it to the various working parts of the engine. The pump is interconnected to the throttle, so that optimum lubrication is achieved at all times, corresponding to the requirements of both engine speed and throttle opening. This system completely obviates the need for petroil and the problems that arise when pre-mixing petrol and oil.

2 Operations with engine in frame

It is not necessary to remove the engine unit from the frame unless the crankshaft assembly and/or the gearbox internals require attention. Most operations can be accomplished with the engine in place, such as:

1 Removal and replacement of the cylinder heads.
2 Removal and replacement of the cylinder barrels and pistons.
3 Removal and replacement of the generator.
4 Removal and replacement of the clutch.
5 Removal and replacement of the contact breaker assembly.
When several operations need to be undertaken simultaneously, it will probably be advantageous to remove the complete engine unit from the frame, an operation that should take approximately two hours, working at a leisurely pace. This will give the advantage of better access and more working space.

3 Operations with engine removed

1 Removal and replacement of the crankshaft assembly.
2 Removal and replacement of the gear cluster, selectors and gearbox main bearings.

4 Method of engine/gearbox removal

As mentioned previously, the engine and gearbox are built in unit and it is necessary to remove the unit complete, in order to gain access to components. Separation is accomplished after the engine unit has been removed from the frame and refitting cannot take place until the crankcase has been reassembled. When the crankcase is separated, the gearbox internals will be exposed.

5 Removing the engine/gear unit

1 Place the machine on the centre stand and make sure that it is standing firmly on level ground.
2 Turn the fuel tap to the position marked 'stop'. Disconnect the two fuel pipes where they join the float chamber of each carburettor. The pipes are a push-on fit. If the tank still contains fuel, drain it off by turning the tap to the 'on' position. When the tank is drained, disconnect the rubber pipe connecting both halves of the petrol tank, on the underside. The tank can now be lifted away from the machine; it is retained in position by rubber buffers at the front and rear of the tunnel through the centre that compress around the top frame tube.

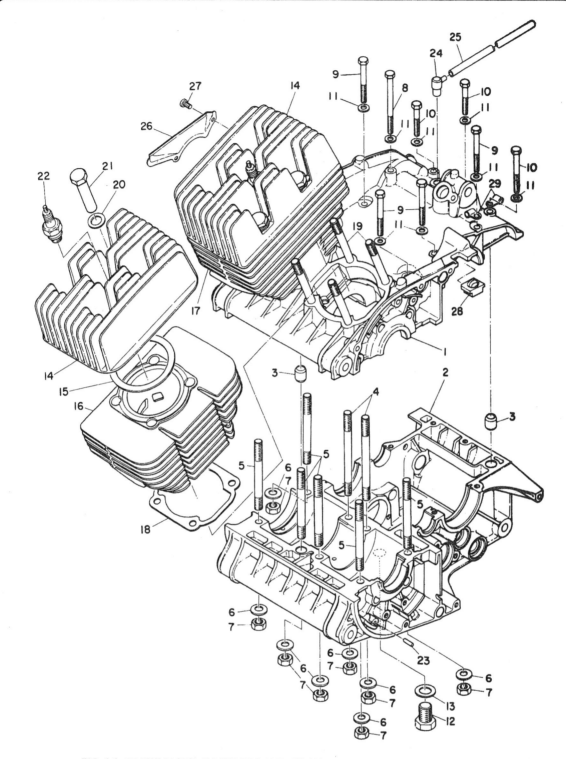

FIG. 1.1. CRANKCASES, CYLINDERS AND CYLINDER HEADS — COMPONENT PARTS

1 Upper crankcase	10 Bolt - 3 off	17 Cylinder barrel - right-hand	22 Spark plug - 2 off
2 Lower crankcase	11 Plain washer - 8 off	18 Cylinder base gasket - 2 off	23 Dowel pin
3 Dowel pin - 2 off	12 Drain plug		24 Breather
4 Crankcase stud - 2 off	13 Drain plug gasket	19 Cylinder holding down stud - 8 off	25 Breather pipe
5 Crankcase stud - 6 off	14 Cylinder head - 2 off		26 Cover
6 Plain washer - 8 off	15 Cylinder head gasket - 2 off	20 Plain washer - 8 off	27 Screw - 2 off
7 Nut - 8 off		21 Holding down nuts - 8 off	28 Clutch cable stop
8 Bolt	16 Cylinder barrel - left hand		29 Oil pipe clamp - 2 off
9 Bolt - 4 off			

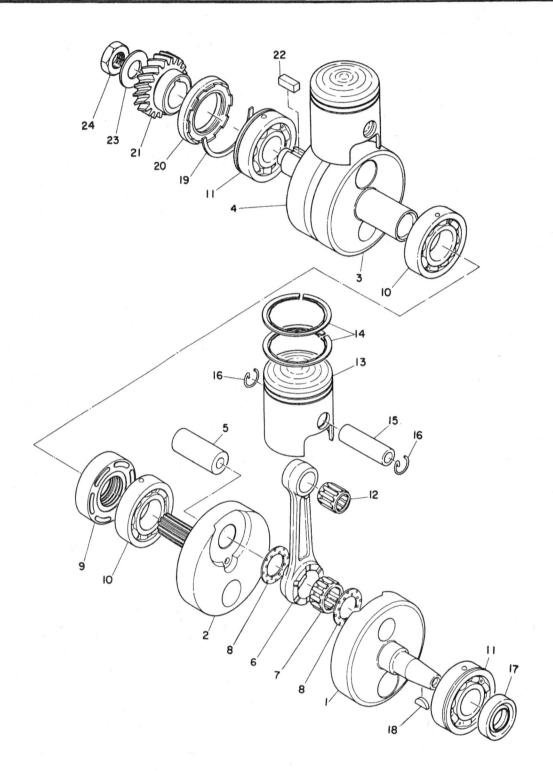

FIG. 1.2. CRANKSHAFT AND PISTONS — COMPONENT PARTS

1 Crankshaft 1 - left-hand	9 Labyrinth seal	17 Oil seal - left-hand
2 Crankshaft 1 - right-hand	10 Main bearing	18 Woodruff key
3 Crankshaft 2 - left-hand	11 Main bearing	19 Circlip
4 Crankshaft 2 - right-hand	12 Needle roller bearing, small end - 2 off	20 Oil seal - right-hand
5 Crankpin - 2 off	13 Piston - 2 off	21 Primary drive pinion (23 teeth)
6 Connecting rod - 2 off	14 Piston ring set	22 Woodruff key
7 Big-end bearing - 2 off	15 Gudgeon pin - 2 off	23 Belville washer
8 Crankpin washer - 4 off	16 Circlip - 4 off	24 Crankshaft nut

5.2 Tank is retained solely by rubbers under compression

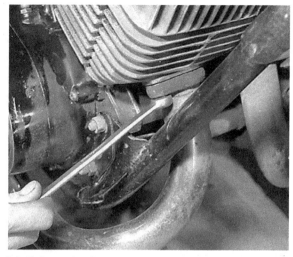

5.4a Exhaust pipe flanges fit over two studs

5.4b Silencer bolts fit below pillion footrests

5.4c Note rubber bush in silencer mounting lug

5.5 Twist exhaust system to free without dismantling footrests

5.6a Oil pump cover is retained by three cross head screws

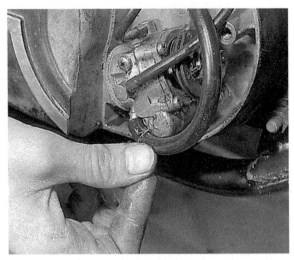

5.6b Spring clip retains feed pipe on oil pump

5.7 Release control cable from pump pulley. Note frayed cable

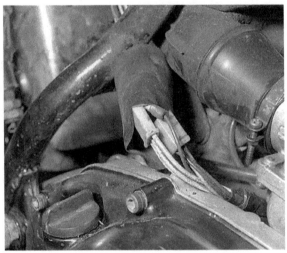

5.8 Electrical snap connectors are within plastic sleeve

5.10a Generator cover is retained by three long cross head screws

5.10b Three additional screws retain crankcase cover

5.10c Small crankcase insert acts as clutch cable stop

5.11a Contact breaker assembly and brush gear comes away with cover

5.11b Lift cover away complete with wiring harness

5.12a Contact breaker cam and rotor are retained by centre bolt

5.12b Contact breaker cam is keyed onto shaft, to preserve correct timing

5.13a Silencer bolt can be used to extract the rotor

5.13b Rotor is keyed onto crankshaft taper

5.14 Detach final drive chain at spring link

5.15a Unscrew clip to release air cleaner hose

5.15b Other end of hose pulls off air cleaner stub

5.16 Tachometer cable is retained by threaded gland nut

5.17 Place throttle valves out of harms way, to prevent damage

5.18a Clip retains carburettor to inlet stub

3 Although there is no necessity to detach the petrol tank in order to remove the engine/gear unit, better access will be gained with less risk of damage to the painted surfaces if the tank is out of the way.

4 Remove the bolts from each exhaust pipe flange at the joint between the exhaust pipe and cylinder barrel. Remove the nuts and bolts that retain each silencer to the lug carrying the pillion footrests. These pass through a rubber bush within the mounting lug of each silencer.

5 It is not necessary to remove the footrests in order to free the exhaust system, even though it may appear there is insufficient clearance. If each pipe is turned through 90º in the outward direction it can be pulled clear without difficulty.

6 Remove the three cross-head screws retaining the front portion of the right-hand crankcase cover; detach the cover. This will give access to the oil pump. Release the wire clip around the oil feed pipe and pull the pipe from the oil pump connection. Allow the oil tank to drain, or alternatively block the end of the pipe with a bolt of convenient size to impede the flow of oil whilst dismantling continues.

7 Disconnect the control cable linking the oil pump with the throttle. A barrel-shaped nipple on the end of the cable engages with the oil pump pulley and can be slipped out of position by twisting whilst the pulley is turned. When the cable is free, remove the adjuster which threads into the outer crankcase cover. The cable can then be withdrawn completely from the engine.

8 The electrical connections are found within a plastic sleeve close to the air intake of the right-hand carburettor. Separate the connectors; note the colour coding.

9 Remove the gear change lever on the left-hand side of the machine. It is retained on a splined shaft by a pinch bolt; when the pinch bolt is slackened, the lever can be pulled off the splines. Mark both the lever and the splined shaft, the lever can then be replaced in the same position.

10 Remove the three cross-head screws securing the generator cover on the left hand side of the machine; remove the cover noting that it is located on dowels and is therefore a tight fit. The outer crankcase cover, also secured by three cross-head screws, can now be detached. Before the crankcase cover can be lifted clear it will be necessary to detach the clutch cable, first from the handlebar lever and then from the clutch operating arm within the cover. Note how the lower end of the cable fits into an abutment attached to a small portion of the crankcase, retained in position by the outer cover.

11 Withdraw the contact breaker assembly complete with the brush gear cover enclosing the generator. This is retained by three cross head screws around the periphery. The assembly can be withdrawn complete with wiring harness if the rubber grommet is detached from the back of the upper crankcase casting. The circular aperture will provide sufficient clearance for the cable connectors to pass through. Note that the separate lead to the neutral contact must lie separated at the snap connector.

12 Detach the contact breaker cam from the generator armature. When the centre retaining bolt is unscrewed, the cam is freed and can be lifted away. It is located by means of a peg and slot arrangement, so that it will always be replaced in the correct position.

13 A Yamaha service tool is recommended to remove the generator armature from the end of the crankshaft. This takes the form of a puller bolt that will release the armature from the tapered end of the crankshaft, when it is screwed home. If the service tool is not available, the same effect can be achieved by using one of the silencer retaining bolts in its place. The taper joint is keyed; remove the key and place it in a safe place for subsequent reassembly.

14 Part the final drive chain at the spring connecting link and remove the chain for later inspection and lubrication.

15 Detach the air cleaner hoses from both carburettors. The hoses are retained around the carburettor bell mouth by a screw clip. Remove both hoses; the other end is a push fit over the outlets from the air cleaner assembly.

16 Disconnect the tachometer drive cable by unscrewing the gland nut. This is found to the rear of both carburettors, close to

the top engine mounting bolt.

17 Unscrew both carburettor tops and remove them, complete with their needle and throttle valve assembly. These parts are very easily damaged and should be taped out of harms way to some other part of the machine.

18 The carburettors can be pulled away from their flexible mountings if each clip fitting is unscrewed first. Slacken the cross head screw that passes through each clip.

19 Remove both spark plug leads by pulling off the plug caps.

20 Remove the four engine mounting bolts and the rear mounting plates attached to the lug on the upper crankcase, close to the tachometer drive take-off. The engine unit can now be lifted from the frame, preferably from the right-hand side. Although the engine unit is not unduly heavy, it is advisable to have a second person available during this operation, if only to steady the machine and ensure there is sufficient clearance as the engine is manipulated out of position.

6 Dismantling the engine and gearbox: general

1 Before commencing work on the engine unit, the external surfaces should be cleaned thoroughly. A motor cycle engine has very little protection from road grit and other foreign matter, which will find its way into the dismantled engine if this simple precaution is not taken. One of the proprietary cleaning compounds, such as "Gunk" or "Jizer" can be used to good effect, particularly if the compound is permitted to work into the film of oil and grease before it is washed away. Special care is necessary when washing down to prevent water from entering the now exposed parts of the engine unit.

2 Never use undue force to remove any stubborn part unless specific mention is made of this requirement. There is invariably good reason why a part is difficult to remove, often because the dismantling operation has been tackled in the wrong sequence. Dismantling will be made easier if a simple engine stand is constructed to correspond with the engine mounting points. This arrangement will permit the complete engine unit to be clamped rigidly to the workbench, leaving both hands free.

7 Dismantling the engine unit: removing the cylinder heads, barrels and pistons

1 Each cylinder head is retained by four sleeve bolts. Slacken them in a diagonal sequence and then remove the cylinder heads, complete with gaskets.

2 Remove the oil delivery pipe clamps from the rear of each cylinder barrel and the banjo unions which connect each pipe to the inlet port. Withdraw the cylinders from the crankcase, one at a time, by lifting them upwards along the holding down studs. Take care to support each piston and rings as it emerges from the cylinder bore; it is a wise precaution to place some clean rag in the mouth of each crankcase before the piston is released from the cylinder so that in the event of piston ring breakages. particles of broken ring will not fall into the crankcase.

3 Remove the circlips from each piston and press out the gudgeon pins so that the pistons are released. If the gudgeon pins are a particularly tight fit, the pistons should be warmed first, to expand the alloy and release the grip on the steel pins. If it is necessary to tap the gudgeon pins out of position, make sure the connecting rod is supported to prevent distortion. On no account use excess force. Discard the old circlips.

4 Before the pistons are placed aside for examination at a later stage, mark each **inside** the skirt so that it is replaced in an identical position. It is advisable to keep the gudgeon pins 'matched' with the pistons from which they were originally extracted. There is no need to mark the back and front of each piston as the front is denoted by an arrow on the piston crown.

5.18b Note flexible rubber mountings for carburettors

5.20a Lower engine bolt is at rear of engine unit

5.20b Upper rear engine bolt passes through short engine plates

5.20c Slide engine unit out from right-hand side

8 Dismantling the engine unit: removing the clutch assembly

1 Lay the crankcase assembly on its left-hand side and remove the right-hand crankcase cover. Before the cover can be removed, it will first be necessary to detach the kickstarter, which is retained on a splined shaft by a pinch bolt. Slacken the pinch bolt and draw the kickstarter off the splines.

2 The crankcase cover is held in position by eight cross head screws and two dowels. When the screws are removed, the cover will pull away, complete with the oil pump after the oil pump cable adjuster is unscrewed from the top. There is no necessity to detach the oil pump as a separate unit, unless replacement is necessary.

3 The clutch assembly is now exposed. Commence by unscrewing the six cross head screws in the pressure plate and remove them, together with the six clutch springs. If the pressure plate is lifted away, the clutch plates can be withdrawn. There are thirteen of these, seven plain plates and six with friction inserts, alternatively spaced. Note the presence of the six rubber cushion rings. Remove the mushroom-headed portion of the clutch push rod assembly which protrudes from the hollow mainshaft.

4 The clutch centre is retained by a very tight nut. The

use of Yamaha service tool YR1 is advised to hold the clutch centre firmly whilst the centre nut is slackened by a socket spanner. ON NO ACCOUNT ATTEMPT TO LOCK THE CLUTCH CENTRE WITH THE OUTER DRUM SERRATIONS, BOTH DRUMS ARE CAST IN LIGHT ALLOY AND WILL FRACTURE VERY EASILY IF OVERSTRESSED IN THIS MANNER. Alternatively, engage bottom gear and hold the final drive sprocket retaining nut so that the mainshaft is locked in position.

5 When the centre retaining nut has been slackened and removed, the clutch centre will pull off the mainshaft splines, followed by a thrust washer. The clutch outer drum and integral primary drive gear will also pull off. Remove the spacer that fits within the clutch outer drum assembly and the thrust washer that precedes the mainshaft oil seal.

9 Dismantling the engine unit: removing the kickstarter assembly

1 Remove the circlip retaining the kickstarter drive pinion; withdraw the pinion from its shaft.

7.1a Cylinder heads are retained by four sleeve bolts

7.1b Heads will lift off after bolts are withdrawn

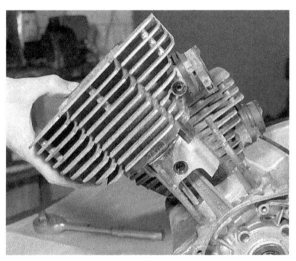

7.2 Slide cylinder barrels up holding down studs

7.3a Remove and discard old circlips

7.3b Warming pistons makes gudgeon pin removal easy

8.2 Right-hand crankcase cover is retained by eight screws

8.3a Remove screws to release pressure plate

8.3b Clutch has six springs

8.3c Lift off pressure plate to expose clutch plates

8.3d Lift out clutch plates and ....

8.3e ... withdraw mushroom headed push rod

8.4 Engage gear and lock engine to release clutch centre retaining nut

8.5a Plain thrust washer is located beneath clutch centre

8.5b Lift off clutch outer drum and primary drive

8.5c Spacer fits over mainshaft, in centre of clutch

8.5d Stepped thrust washer is located in front of main bearing

9.1 Circlip retains kickstarter idler pinion

10.2 Lock engine to remove primary drive pinion from crankshaft

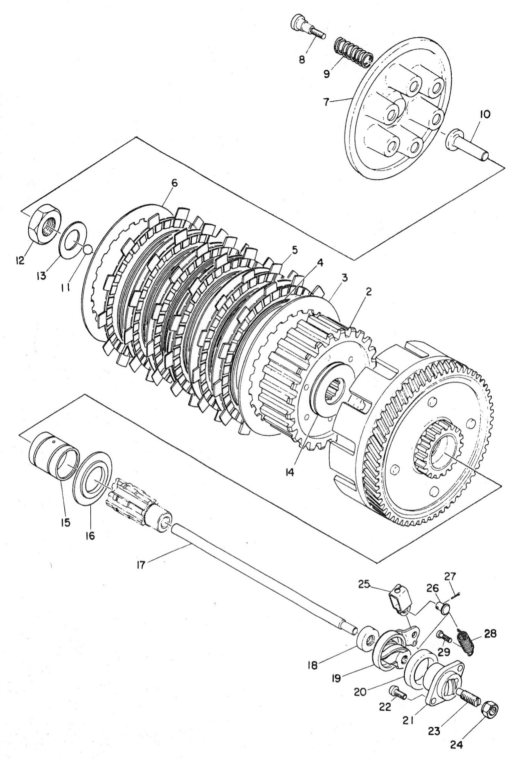

FIG. 1.3. CLUTCH — COMPONENT PARTS

1 Primary driven gear and clutch outer drum
2 Clutch inner drum
3 Plain clutch plate
4 Cushion ring - 6 off
5 Friction (inserted) clutch plate - 6 off
6 Plain clutch plate - 6 off
7 Pressure plate
8 Clutch spring screw - 8 off
9 Clutch spring - 8 off
10 Clutch push rod 'mushroom'
11 Ball bearing (5/16 in. diam.)
12 Lock nut
13 Belville spring washer
14 Thrust washer
15 Spacer
16 Thrust washer
17 Push rod
18 Push rod seal
19 Clutch actuating mechanism
20 Dust seal
21 Actuating mechanism housing
22 Screw - 2 off
23 Adjusting screw
24 Adjusting nut
25 Trunnion for cable
26 Clevis pin
27 Split pin
28 Return spring
29 Spring anchor

2 The kickstarter ratchet assembly is retained by the kickstarter return spring, one end of which is looped around a projection cast into the crankcase. Remove the spring and then the kickstarter assembly as a complete unit. There is no necessity to dismantle the unit further, unless replacements are required. See Section 24 of this Chapter for further details.

10 Dismantling the engine unit: removing the primary drive pinion

1 Lock the engine in position by placing a stout metal rod through the eye of both connecting rods, resting it on two supports across the crankcase.
2 Unscrew and remove the pinion retaining nut (right-hand thread) and pull the pinion off the end of the crankshaft. It is located by means of a key, which should be removed and placed in a safe place, together with the domed washer between the retaining nut and the pinion. No difficulty should be experienced in removing the pinion because it is a parallel fit on the mainshaft.

11 Dismantling the engine unit: removing the gear change mechanism

1 Replace the engine in the upright position, resting on the bottom of the crankcase. Remove the rubber cover over the gear change shaft on the left-hand side and the circlip and shim washer that lie beneath. The complete shaft can now be pulled out from the right-hand side of the engine unit. A slot in the arm attached to the end of the shaft engages with a peg on the operating arm of the gear change mechanism. Do not lose the shouldered collar that slides over the peg.
2 Remove the circlip from the gear selector operating arm and lift the arm away from the spindle, on which it pivots. It will be necessary to open out the two spring-loaded pawls so that they will disengage from the pins of gear selector drum. The operating arm will come away complete with the return spring.

12 Dismantling the engine unit: removing the final drive sprocket

1 Working from the left-hand side of the engine unit, hold the final drive sprocket with the appropriate Yamaha service tool or alternatively wrap the final drive chain around the sprocket and secure both ends in a vice so that the sprocket is prevented from moving.
2 Bend back the tab washer that locks the sprocket retaining nut and unscrew the nut. It is best to use a good fitting socket spanner for this purpose since the nut is very tight and may require considerable leverage to move it initially. When the nut is removed the sprocket will pull off the shaft without difficulty.

13 Dismantling the engine unit: separating the crankcase

1 Invert the crankcase and remove the eight nuts (13 mm spanner) each of which is numbered. The nuts should be slackened in numerical order, commencing with the highest number and working downward.
2 When all the bottom nuts and washers have been detached, turn the crankcase upright again and remove the eight bolts (10 mm spanner) on the top surface. These too are numbered and should be slackened and removed in similar order.
3 When all sixteen nuts and bolts have been withdrawn, the crankcases can be separated. Part them by striking the front part of the upper crankcase and the rear part of the lower crankcase with a soft-faced mallet. Tap gently - little force is needed to effect the separation.

11.2a Circlip retains gear selector operating arm

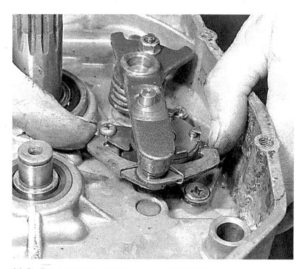

11.2b Pawls must be sprung apart whilst lifting arm off pivot

12.2 Final drive sprocket fits splines of layshaft

13.3a Crankcases separate horizontally after bolts are withdrawn

13.3b Complete assembly remains in lower crankcase

14.1 Lift crankshaft assembly from lower crankcase

14.2a Lift out gear clusters complete, first mainshaft ...

14.2b ... then layshaft

14.3a Selector forks and rods remain in position as shown

14.3b Oil seals must be removed to release selector rods

14.3c Selector forks are sliding fit on selector rods

15.2 Plastic tachometer drive pinion is retained by circlip

14 Dismantling the engine unit: removing the crankshaft assembly and gear clusters

1 Remove the crankshaft by tapping each end with a soft-faced mallet. It will come away as a complete unit, together with the three oil seals, one at each end of the crankshaft and one in the centre, between the two centre ball races.

2 The gear clusters will lift out in similar fashion, together with their respective shafts, bearings and oil seals.

3 The selector rods on which the selector forks pivot are a sliding fit in the lower crankcase and can be pushed out from the right-hand side. Both have an oil seal on the left-hand side, which must be removed before the selector rods can be freed. A circlip at the left-hand end of each rod prevents their displacement from the left. The selector rod carrying the two selector forks has a circlip inside the crankcase at the left-hand end.

4 When the selector rods have been removed, the selector forks will be free to lift away since the rods pass through them. Mark the forks so that they are replaced in the same positions; the forward facing selector rod carries a pair of 'handed' forks and the rear facing rod, only a single fork.

5 Remove the circlip halves used to locate the various ball races in the engine and gearbox assembly, and place them aside until reassembly commences.

15 Dismantling the engine unit: removing the neutral switch, tachometer drive pinion and oil pump drive pinion

1 There is no necessity to remove the neutral indicator switch unless the switch malfunctions or the gear selector drum is to be removed. The switch unit is retained to the crankcase by three cross head screws which, when withdrawn, will release the switch unit.

2 If the gear selector drum is to be removed, unscrew the cross head screw in the centre of the end of the drum and detach the circular end plate together with the neutral contact and spring.

3 The tachometer drive pinion, made of plastic material, is retained by a circlip. Remove the circlip and the flat washer behind it, then lift off the pinion. Note that a small metal rod passes through a drilling in the end of the drive shaft, to line up with a slot in the pinion centre. When the pinion is removed, withdraw the rod and keep it in a safe place for reassembly at a withdraw the rod and keep it in a safe plate for reassembly.

4 The oil pump drive pinion is made of the same plastic material and is secured to the oil pump drive spindle by means of a nut and shakeproof washer. When the nut and washer are removed, the pinion can be pulled off the spindle.

5 It is necessary to remove the pinions only if damage has occurred. Although damage rarely occurs, the occasional chipped or broken tooth may be found when the engine is dismantled.

16 Dismantling the engine unit: removing the gear selector drum

1 It is unlikely that there will be any need to remove the gear selector drum from the lower crankcase unless the tracks followed by the gear selector forks are damaged. Commence by removing the two cross head screws that retain the gear change lever guide and the guide itself.

2 Remove the two countersunk cross head screws that hold the retaining plate for the gear selector drum. Remove the plate.

3 Unscrew and remove the plunger assembly from the base of the crankcase. The hollow bolt that holds the plunger and spring threads into the rear of the lower crankcase at an angle, on the left-hand side.

4 Press the gear selector drum out of its housing from the left-hand side until the circlip around the left-hand end of the drum can be removed. This will free the cam on the end of the drum, which can be pulled away. The remainder of the drum will now pass through the aperture in the right-hand end of the crankcase, permitting the drum to be removed completely.

16.2 Retaining plate holds selector drum in position

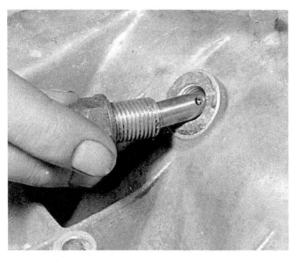

16.3 Plunger holder unscrews from crankcase underside

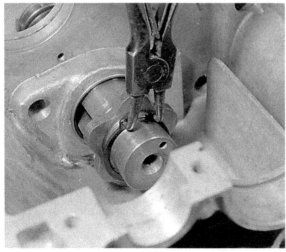

16.4 Circlip retains cam on end of selector drum

17 Dismantling the engine unit: removing the tachometer drive

1   Remove the tachometer drive pinion (Section 15.3). This will give access to the spindle retaining plate, which is secured to the crankcase casting by three cross head screws. Remove the three screws and the retaining plate. Pull out the drive spindle from the right, after removing the circlips which hold the worm drive pinion on the spindle.

2   The driven spindle passes through a plate attached to the top of the upper crankcase by a bolt. Remove the bolt and withdraw the driven spindle assembly complete.

18 Examination and renovation: general

1   Before examining the component parts of the dismantled engine/gear unit for wear, it is essential that they should be cleaned thoroughly. Use a paraffin/petrol mix to remove all traces of oil and sludge which may have accumulated within the engine.

2   Examine the crankcase castings for cracks or other signs of damage. If a crack is discovered, it will require professional attention, or in an extreme case, renewal of the casting.

3   Examine carefully each part to determine the extent of wear. If in doubt, check with the tolerance figures whenever they are quoted in the text. The following sections will indicate what type of wear can be expected and in many cases, the acceptable limits.

4   Use clean, lint-free rags for cleaning and drying the various components, otherwise there is risk of small particles obstructing the internal oilways.

19 Crankshaft assembly: examination and replacement

1   The crankshaft assembly comprises two separate sets of flywheels with their respective big ends, connecting rods and main bearings, pressed together to form a single unit. A special oil seal of the labyrinth type is interposed between the two inner main bearings of each flywheel set.

2   In the event of main bearing failure, it is beyond the means of the average owner to separate the flywheel assemblies and to realign them to the high standard of accuracy required. In consequence, the complete crankshaft assembly must be taken to a Yamaha specialist for the necessary repairs and renovation, or service exchanged for a fully reconditioned unit.

3   Main bearing failure will immediately be obvious when the bearings are inspected after the old oil has been washed out. If any play is evident or if the bearings do not run freely, renewal is essential. Warning of main bearing failure is usually given by a characteristic rumble that can be readily heard when the engine is running. Some vibration will also be felt, which is transmitted via the footrests.

4   Big end failure is characterised by a pronounced knock that will be most noticeable when the engine is working hard. There should be no play whatsoever in either of the connecting rods, when they are pushed and pulled in a vertical direction. A small amount of sideways play is permissible, but not more than 0.3 mm (0.012").

5   Oil seal failure is a common occurrence in two-stroke engines that have seen a reasonable amount of service. When the oil seals begin to wear, air is admitted to the crankcase which will dilute the incoming mixture whilst it is under crankcase compression. This in turn causes uneven running and difficulty in starting.

6   The oil seals at each end of the crankshaft are easy to renew when the engine is stripped; they are a push fit over each end of the crankshaft, one against and the other close to the outer main bearings. It is a wise precaution to renew these seals whenever the engine is stripped, irrespective of their condition.

7   The labyrinth oil seal in the centre of the crankshaft assembly can be removed and replaced only when the crankshaft assembly is separated. It is difficult to give any guide lines about the need for replacement of this seal as it normally has a long working life, akin to that of the crankshaft assembly itself. If the

17.1 Release circlips to free pinion from shaft

28.2a Forward-facing selector rod carries single fork

28.2b Note inner circlip on rod carrying two selector forks

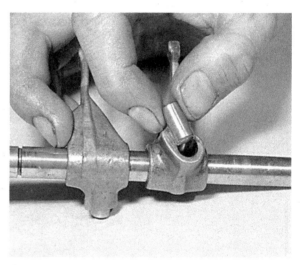

28.3 Pins in selector forks are loose fit

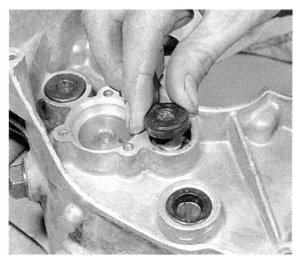

28.5 Always fit new oil seals

28.7 Gear change operating arm is retained by circlip

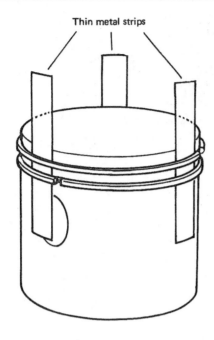

Thin metal strips

Fig. 1.4. Freeing gummed rings

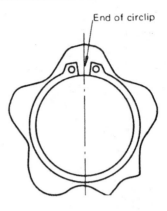

End of circlip

Fig. 1.5. Correct location of selector drum circlip

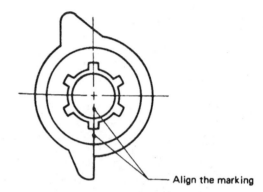

Align the marking

Fig. 1.7. Correct alignment of ratchet wheel clip

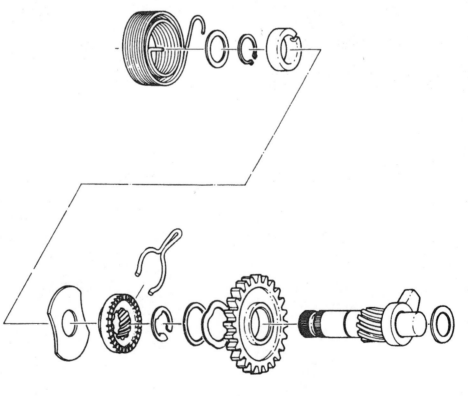

Fig. 1.6. Kickstarter shaft assembly — component parts

qualities of the middle seal are suspect, check whether the seal spins quite freely on the crankshaft. If it does, the chances are that the seal is due for renewal; it will be supplied as part of the built-up crankshaft assembly, after the Yamaha repair specialist has separated the two sets of flywheels to remove the old oil seal and re-aligned them after fitting the new seal.

20 Small end bearings: examination and replacement

1 The small end bearings are caged needle rollers and will seldom give trouble unless a lubrication failure has occurred. The gudgeon pins should be a good sliding fit in their bearings, without any play. If play develops, a noticeable rattle will be heard when the engine is running, indicative of the need for renewal of the bearings.

2 No problem is encountered when replacing the caged needle roller bearings as they are a light push fit in the eye of each connecting rod. New small end bearings are normally supplied whenever the crankshaft assembly is renewed or service-exchanged.

21 Pistons and piston rings: examination and renovation

1 Attention to the pistons and rings can be overlooked if a rebore is necessary because new pistons and rings will be fitted. under these circumstances.

2 If a rebore is not considered necessary, each piston should be examined closely. Reject the piston if it is badly scored or dis-coloured as the result of the exhaust gases by-passing the rings. Check the gudgeon pin bosses to ensure that they are not enlarged or that the grooves retaining each circlip are not damaged.

3 Remove all carbon from each piston crown and use metal polish to finish off, so that a high polish is obtained. Carbon will adhere much less readily to a polished surface. Examination of the piston crown will show whether the engine has been rebored previously, since the amount of overbore is invariably stamped on each piston crown. Two oversizes are available + 0.25 mm and + 0.50 mm.

4 The grooves in which the piston rings locate can become enlarged in use. The clearance between the edge of each piston ring and the groove in which it seats should not exceed 0.002″.

5 Remove the piston rings by pushing the ends apart with the thumbs whilst gently easing the ring from its groove. Great care is necessary throughout this operation because the rings are brittle and will break easily if overstressed. If the rings are gummed in their grooves, three strips of tin can be used to ease them free, as shown in the accompanying illustration.

6 Piston ring wear can be checked by inserting the rings one at a time in the cylinder bore from the top and pushing them down about 1½ inches with the base of the piston so that they rest square in the bore. Make sure that the end gap is away from any of the ports. If the end gap is within the range 0.45 - 0.65 mm (0.017″ - 0.025″) the ring is fit for further service.

7 Examine the working surface of each piston ring. If dis-coloured areas are evident, the ring should be renewed because these areas indicate the blow-by of gas. Check that there is not a build-up of carbon on the back of the ring or in the piston ring groove, which may cause an increase in the radial pressure. A portion of broken ring affords the best means of cleaning out the piston ring grooves.

8 Check that the piston ring pegs are firmly embedded in each piston ring groove. It is imperative that these retainers should not work loose, otherwise the rings will be free to rotate and there is danger of the ends being trapped in the ports.

9 It cannot be over-emphasised that the condition of the pistons and piston rings is of prime importance because they control the opening and closing of the ports by providing an effective moving seal. A two-stroke engine has only three working parts, of which the piston is one. It follows that the efficiency of the engine is very dependent on the condition of pistons and the parts with which they are closely associated.

22 Cylinder barrels: examination and renovation

1 There will probably be a lip at the uppermost end of each cylinder barrel which marks the limit of travel of the top of the piston ring. The depth of the lip will give some indication of the amount of bore wear that has taken place even though the amount of wear is not evenly distributed.

2 Insert each piston (without rings) in turn in its respectively cylinder bore so that it is about ¾″ from the top of the bore. Measure the clearance between the piston skirt and the cylinder wall with a feeler gauge. Repeat at two further positions lower down the bore. The recommended clearance is from 0.040 - 0.045 mm (0.0016″ - 0.0018″); if the clearance exceeds 0.050 mm (0. 0019″) the cylinder is in need of a rebore. Always rebore the cylinders as a pair and fit oversize pistons.

3 Give the cylinder barrels a close visual inspection. If the surface of either bore is scored or grooved, indicative of an earlier seizure or a displaced circlip and/or gudgeon pin, rebore is essential. Compression loss has a marked effect on engine performance.

4 Check that the outside of each cylinder barrel is clean and free from road dirt. Use a wire brush on the cooling fins if they are obstructed in any way but take care not to score or damage the light alloy. If the air flow to the cooling fins is obstructed, the engine may overheat badly. Although caustic soda is often recommended for cleaning some of the oilier components of a two-stroke engine, NEVER use on parts made of light alloy. Caustic soda attacks aluminium alloy with great vigour and produce an explosive gas.

5 Clean all carbon deposits from the exhaust ports, using a blunt ended scraper. It is important that all the ports should have a clean, smooth appearance because this will have the dual benefit of improving gas flow and making it less easy for carbon to adhere in the future. Finish off with metal polish, to heighten the polishing effect.

6 Do not under any circumstances enlarge or alter the shape of the ports, under the mistaken belief that improved performance will result. The size and position of the ports predetermines the characteristics of the engine and unwarranted tampering can produce very adverse effects.

23 Cylinder heads: examination and renovation

1 It is unlikely that the cylinder heads will require any special attention apart from removing the carbon deposit from the combustion chamber. Finish off with metal polish; the polished surface will help improve gas flow and reduce the tendency of future carbon deposits to adhere so readily. .

2 Check that the cooling fins are clean and unobstructed, so that they receive the full air flow. If a wire brush is used, take care not to scratch or damage the fins.

3 Check the condition of the thread within each spark plug hole. The thread is easily damaged if the spark plug is over-tightened. If necessary, a damaged thread can be reclaimed by fitting a Helicoil thread insert. Most Yamaha agents have facilities for this type of repair, which is not expensive.

4 If there has been evidence of oil seepage from the cylinder head joint when the machine was in use, check whether either cylinder head is distorted by laying it on a sheet of plate glass. Severe distortion will necessitate renewal of the cylinder head, but if the distortion is only slight, the head can be re-claimed by wrapping a sheet of emery cloth around the glass and using it as the surface on which to rub down the head with a rotary motion, until it is once again flat. The usual cause of distortion is failure to tighten down the cylinder head bolts evenly in a diagonal sequence.

24 Gearbox components: examination and renovation

1 Examine each of the gear pinions to ensure that there are no chipped or broken teeth and that the dogs on the end of the pinions are not rounded. Gear pinions with any of these defects

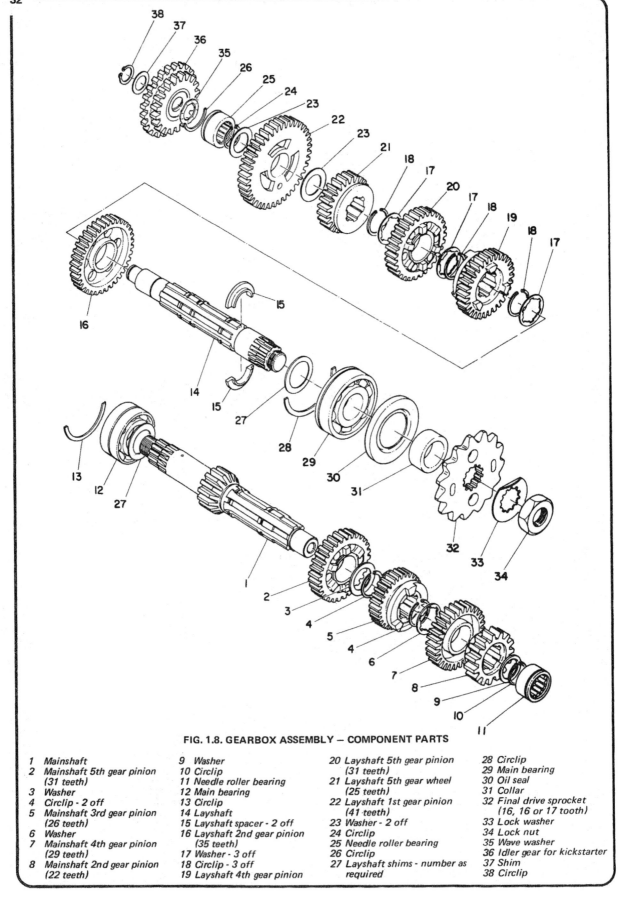

FIG. 1.8. GEARBOX ASSEMBLY — COMPONENT PARTS

1 Mainshaft	9 Washer	20 Layshaft 5th gear pinion	28 Circlip
2 Mainshaft 5th gear pinion	10 Circlip	(31 teeth)	29 Main bearing
(31 teeth)	11 Needle roller bearing	21 Layshaft 5th gear wheel	30 Oil seal
3 Washer	12 Main bearing	(25 teeth)	31 Collar
4 Circlip - 2 off	13 Circlip	22 Layshaft 1st gear pinion	32 Final drive sprocket
5 Mainshaft 3rd gear pinion	14 Layshaft	(41 teeth)	(16, 16 or 17 tooth)
(26 teeth)	15 Layshaft spacer - 2 off	23 Washer - 2 off	33 Lock washer
6 Washer	16 Layshaft 2nd gear pinion	24 Circlip	34 Lock nut
7 Mainshaft 4th gear pinion	(35 teeth)	25 Needle roller bearing	35 Wave washer
(29 teeth)	17 Washer - 3 off	26 Circlip	36 Idler gear for kickstarter
8 Mainshaft 2nd gear pinion	18 Circlip - 3 off	27 Layshaft shims - number as	37 Shim
(22 teeth)	19 Layshaft 4th gear pinion	required	38 Circlip

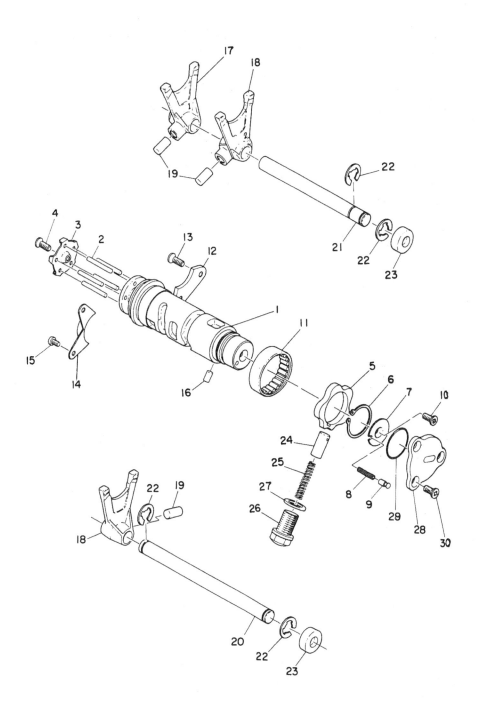

FIG. 1.9. GEAR SELECTOR MECHANISM – COMPONENT PARTS

1 Gear selector drum
2 Gear selector drum plate pins - 5 off
3 Gear selector plate
4 Screw
5 Stopper plate
6 Circlip
7 Neutral contact plate
8 Spring
9 Neutral contact
10 Screw

11 Needle roller bearing
12 Retaining plate
13 Screw - 2 off
14 Change lever guide
15 Screw - 2 off
16 Dowel pin
17 Gear selector fork 1
18 Gear selector fork 2 - 2 off
19 Cam follower pin - 3 off
20 Shaft for gear selector forks 1 and 2

21 Shaft for gear selector fork 2
22 Circlip - 4 off
23 Blanking off plug - 2 off
24 Cam stopper
25 Cam stopper spring
26 Cam stopper plug
27 Gasket for cam stopper plug
28 Neutral switch assembly
29 'O' ring seal
30 Screw - 3 off

must be renewed; there is no satisfactory method of reclaiming them.

2 The gearbox bearings must be free from play and show no signs of roughness when they are rotated. Each shaft has a ball journal bearing at one end and a caged needle roller bearing at the other.

3 It is advisable to renew the gearbox oil seals irrespective of their condition. Should a re-used oil seal fail at a later date, a considerable amount of dismantling is necessary to gain access and renew it.

4 Check the gear selector rods for straightness by rolling them on a sheet of plate glass. A bent rod will cause difficulty in selecting gears and will make the gear change action particularly heavy.

5 The selector forks should be examined closely, to ensure that they are not bent or badly worn. Wear is unlikely to occur unless the gearbox has been run for a period with a particularly low oil content.

6 The tracks in the gear selector drum, with which the selector forks engage, should not show any undue signs of wear unless neglect has led to under lubrication of the gearbox. Check that the plunger spring bearing on the cam plate plunger has not lost its action and that the springs of the gear change lever pawl assembly have good tension. Any damage to, or weakness of, the gear change lever return spring will be self-evident.

7 If the kickstarter has shown a tendency to slip, or if it is desired to inspect the kickstarter ratchet, the kickstarter assembly must first be dismantled. Remove the circlip from the splined end of the kickstarter shaft and then withdraw the spacer, spring cover and spring clip around the ratchet wheel. Any wear of the ratchet teeth will be self-evident and it will be necessary to renew both the ratchet wheel and the kickstarter gear pinion with which it makes contact. Examine the kickstarter return spring, which should also be renewed if there is any doubt about its condition.

8 As mentioned earlier in the text, instances have occurred where a tooth has either chipped or broken on the pinions that drive the oil pump and the tachometer drive cable take-off. These pinions are made of plastic material and must be renewed when damage of this nature is found.

25 Clutch assembly: examination and renovation

1 After an extended period of service, the clutch linings will wear and promote clutch slip. The limit of wear measured across each inserted plate is 0.27 mm (0.107''). When the overall width of the linings reach this level, the clutch plates must be replaced as a complete set.

2 The plain clutch plates should not show any evidence of excess heating (bluing) and should not be more than 0.005'' out of true.

3 The clutch springs should have a free (uncompressed) length of 36 mm. If the springs have taken a set of 1 mm or more, the complete set must be renewed.

4 A worn clutch spacer is responsible for clutch noise and should be renewed if the fit within the clutch centre is particularly slack. Check the inner and outer surfaces for scratches; these will impair clutch action if not smoothed away.

5 Check the condition of the slots in the outer surface of the clutch centre and the inner surfaces of the outer drum. In an extreme case, clutch chatter may have caused the tongues of the inserted plates to make indentions in the slots of the outer drum, or the tongues of the plain plates to indent the slots of the clutch centre. These indentations will trap the clutch plates as they are freed and impair clutch action. If the damage is only slight the indentations can be removed by careful work with a file and the burrs removed from the tongues of the clutch plates in similar fashion. More extensive damage will necessitate renewal of the parts concerned.

6 The clutch release mechanism attached to the inside of the left-hand crankcase cover does not normally require attention, provided it is greased at regular intervals. It is held to the cover by two cross head screws and operates on the worm and quick start thread principle. A light return spring ensures that the pressure is taken from the end of the clutch push rod when the handlebar lever is released and the clutch fully engaged.

26 Crankcase covers: examination and renovation

1 The right and left-hand crankcase covers are unlikely to suffer damage unless the machine is dropped or damaged in an accident. If the right-hand cover is fractured, the kickstarter will be rendered inoperative since the cover acts as the outer bearing for the kickstarter shaft. Renewal is therefore essential.

2 The covers have a matt black anodised finish, accentuated by raised portions of the casting that have a polished surface. Small chips can be repaired by over-painting with a matt finish such as 'cylinder black'. This is only a temporary expedient. To fully restore the original finish it will be necessary to have the outer surface of the cover shotblasted and then re-anodised, or alternatively to renew the cover.

27 Engine reassembly - general

1 Before reassembly of the engine/gear unit is commenced, the various component parts should be cleaned thoroughly and placed on a sheet of clean paper, close to the working area.

2 Make sure all traces of old gaskets have been removed and that the mating surfaces are clean and undamaged. One of the best ways to remove old gasket cement is to apply a rag soaked in methylated spirit. This acts as a solvent and will ensure that the cement is removed without resort to scraping and the consequent risk of damage.

3 Gather together all of the necessary tools and have available an oil can filled with clean engine oil. Make sure all new gaskets and oil seals are to hand, also all replacement parts required. Nothing is more frustrating than having to stop in the middle of a reassembly sequence because a vital gasket or replacement has been overlooked.

4 Make sure that the reassembly area is clean and that there is adequate working space. Refer to the torque and clearance settings wherever they are given. Many of the smaller bolts are easily sheared if over-tightened. Always use the correct size screwdriver bit for the cross head screws and never an ordinary screwdriver or punch. If the existing screws show evidence of maltreatment in the past, it is advisable to renew them as a complete set.

28 Engine reassembly: replacing the gear selector mechanism

1 Engine assembly commences with the lower crankcase, which should be positioned in the centre of the workbench. If the gear selector drum has been removed, this must be replaced first by reversing the dismantling procedure given in Section 16. It is important that the circlip retaining the camplate at the end of the drum is located in the correct position. The accompanying illustration shows how this must be aligned.

2 Insert the gear slector rods from the left-hand side of the engine, one at a time. The forward-facing rod passes through a single selector fork, which engages with the right-hand straight cut track of the selector drum. The rod in the rearmost position carries two selector forks, the right-hand one of which engages with the right-hand track (straight) of the selector drum and the left-hand one with the left-hand (curved) track.

3 The pins which engage with the selector drum tracks are a push fit in the selector forks. Make sure that they are not misplaced during assembly as they may fall out of position when the gearbox parts are being moved.

4 There is no necessity to select any particular gear during assembly of the gearbox since the system of gear indexing is quite automatic and does not require 'timing'.

29.2a Half circlips retain main bearings

29.2b Check that selector arms locate correctly at this stage

30.1 Crankshaft will not seat unless knock pins locate

31.3 Tighten crankcase bolts in numerical sequence

31.4 Underside of crankcase has retaining nuts

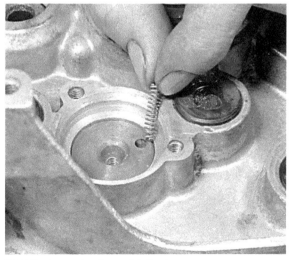

32.1a Replace spring first and follow by ...

32.1b ... circular plate with metal contact

32.1c Tighten down circular plate, then ...

32.2 ... fit plastic end cover

33.1a Left-hand oil seal is positioned as shown

33.1b Right-hand oil seal fits in conventional manner

33.2 Gearbox oil seals must be renewed too

5 When the selector rods and forks are in position, fit new oil seals into the recesses on the left-hand side of the crankcase, they serve the dual purpose of retaining the selector rods in position after the locating circlips.

6 Replace the camplate plunger assembly in the underside of the lower crankcase and check that the end cap is tight and has a new sealing washer.

7 Position the lower crankcase with the right-hand side uppermost and refit the gear selector mechanism. The operating arm fits over a spindle which protrudes from the crankcase casting; it is retained in postion by a circlip. Lubricate the spindle before the operating arm is replaced. It will be necessary to hold apart the two spring-loaded pawls whilst the arm is lowered into position, so that they clear the pins on the end of the gear selector drum.

29 Engine reassembly: replacing the gear clusters

1 Engine reassembly should only be commenced after any dismantled gear train has been reassembled as a complete unit with the gearbox bearings.

2 Position the lower crankcase on its base and insert the three bearing retainers that take the form of half circlips. Lower each complete gear cluster in turn so that the bearings engage with the circlips and the selector fork(s) with the track around the sliding gear pinion. It may be necessary to tap each gear shaft with a soft-faced mallet before they will bed down completely. Check that the shafts revolve quite freely; if not, some part of the assembly is out of alignment.

3 Oil the shafts, gear pinions and bearings so that there will be no dry spots during the eventual start-up of the engine.

30 Engine reassembly: replacing the crankshaft

1 Replace the half circlip bearing retainer in the right-hand bearing housing. Lower the crankshaft assembly into position, noting that each of the main bearings has what is known as a 'knock pin' which engages with a small depression at the leading edge of the crankcase joint.

2 Give both ends of the crankshaft assembly a light tap with a soft-faced mallet, to ensure the assembly is located correctly. Check that each of the knock pins is seating correctly in the bearing housings. Position the crankcase locating dowels in the front and rear of the lower crankcase casting.

31 Engine reassembly: joining the upper and lower crankcase

1 Check that the upper crankcase casting has the tachometer drive assembly fitted. If it is necessary to replace the drive components, reverse the dismantling procedure described in Section 17.

2 Coat the jointing surfaces of both crankcases with a thin layer of gasket cement and lower the upper crankcase casting into position. It may be necessary to give the casting a few light taps with a soft-faced mallet before the jointing surfaces will mate up correctly. DO NOT USE FORCE. If the crankcase will not align, one of the bearing assemblies has not seated correctly.

3 Replace the eight bolts (10 mm spanner) in the top of the crankcase and tighten them in the numerical sequence stamped on the crankcase casting.

4 Invert the crankcase and replace the eight nuts on the underside (13 mm spanner). These too should be tightened in the numerical sequence indicated.

32 Engine reassembly: replacing the neutral indicator

1 If the neutral indicator has been removed from the left-hand end of the gear selector drum it will be necessary first to replace the spring and metal contact, then the circular plate attached to the end of the drum by means of a central, cross head screw. Before the circular plate is tightened down, check that the spring-loaded contact is engaged correctly with the slot in the plate and will move quite freely. The metal contact is waisted to engage with the slot.

2 Replace the end cover; secure by the three countersunk cross head screws. The end cover is made of a plastics material and has an inner 'O' ring seal, which must be in good condition.

33 Engine reassembly: replacing the crankshaft and gearbox oil seals

1 The crankshaft assembly has two oil seals, one of large diameter on the right-hand side and the other of smaller diameter on the left-hand side. The smaller of the two seals is replaced in what may at first appear to be the reverse position, with the spring around the centre boss facing OUTWARD. As a check, the face of the seal is marked to show which is the outer side. The larger seal is fitted in the conventional manner and should have the serrated portion of the face positioned inward, towards the main bearing.

2 The gearbox has three oil seals, two on the left-hand side and one on the right. There is a large diameter seal immediately in front of both the main bearings and a smaller seal in front of the end of the hollow mainshaft, through which the clutch push rod protrudes.

3 Before the oil seals are replaced, the shafts over which they fit should be lubricated thoroughly. Take care to ensure the lips of the seals are not damaged as they are driven into position. On no account use force, otherwise the risk of damage will be high. When positioned correctly, the outer face of each seal should be flush with the edge of the crankcase casting. Before continuing with the reassembly, check that the crankshaft and the gearbox shafts still revolve quite freely.

34 Engine reassembly: replacing the final drive sprocket

1 With the engine resting with its left-hand side uppermost, it is convenient to replace the final drive sprocket. If the original is badly worn and has hooked or broken teeth, it must be renewed at this stage, otherwise a harsh transmission and very rapid chain wear will occur.

2 The sprocket fits over the splined end of the gearbox layshaft and is retained by a large diameter nut and tab washer which also fits over the splines.

3 Hold the sprocket firmly by wrapping the chain around the teeth and securing both ends in a vice, then tighten the centre nut fully (right-hand thread). Bend the tab washer so that it locks the retaining nut in position.

35 Engine reassembly: replacing the primary drive pinion

1 The primary drive pinion is retained on the parallel-shouldered end of the crankshaft by an oblong key and a keyway. Oil the outer surface of the shoulder behind the pinion teeth and slide the pinion on to the crankshaft so that the keyways in both shaft and pinion coincide. Drive the oblong key into position, then add the domed washer and the pinion retaining nut. Assuming that the engine is in gear, with the final drive sprocket still held tightly by the chain clamped in a vice, tighten the pinion retaining nut. Release the chain from the vice.

36 Engine reassembly: replacing the kickstarter mechanism

1 If the kickstarter shaft assembly has been dismantled, it must be reassembled as a complete unit, prior to replacement in the outer crankcase. The accompanying diagram shows the correct

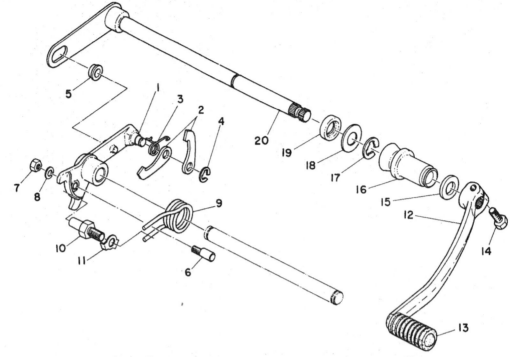

FIG. 1.10. GEAR CHANGE MECHANISM – COMPONENT PARTS

1 Gear change arm	6 Adjusting screw
2 Gear change pawl - 2 off	7 Nut
3 Pawl spring	8 Washer
4 Circlip	9 Return spring
5 Gear change arm roller	10 Stop

11 Lock washer	16 Sealing rubber
12 Gear change lever	17 Circlip
13 Rubber for gear change lever	18 Washer
14 Bolt	19 Oil seal
15 Seal for gear change lever shaft	20 Gear change lever shaft

sequence of assembly.

2   Lubricate the shaft before it is inserted into the crankcase boss and arrange the stop so that it is hard against the crankcase abutment when the shaft is turned in a clockwise direction. Under these circumstances there will be a positive tension on the kickstarter return spring, even when the kickstarter is in the fully-raised position. The end of the spring locates with a peg projection from the casing and should be turned anti-clockwise to locate without moving the shaft. The ratchet wheel clip locates with a cutaway below this peg.

37 Engine reassembly: replacing the tachometer drive pinion

1   Although there is no necessity to remove the drive pinion under normal circumstances (it does not impede dismantling or reassembly) it may be necessary to replace the pinion on the odd occasion.

2   A short metal rod passes through the spindle on which the pinion runs, which should be inserted and spaced so that both ends are equally spaced from the centre. Lower the pinion into position so that the ends of the rod engage with the slots in the centre boss. Replace the washer which fits within the boss, then the circlip that secures the whole assembly.

38 Engine reassembly: replacing the gear change shaft

1   Insert a new oil seal into the lower passageway on the left-hand side of the engine, through which the gear change shaft passes.

2   Insert the gear change shaft from the right-hand side, lubricating the shaft to ease its passage through the oil seal.

3   Slip the shouldered collar, larger diameter downward, over

the peg on the gear selector arm and press home the gear change shaft so that the slotted portion of the operating arm engages with the shouldered collar and peg.

4   Hold the gear change shaft in this position and replace the shim washer and circlip around the left-hand end of the shaft, to retain it in position.

39 Engine reassembly: reassembling and replacing the clutch

1   Position the engine so that the right-hand side faces uppermost and replace the stepped thrust washer over the gearbox mainshaft, immediately in front of the main bearing. The smaller diameter of the washer should face inward.

2   Replace the kickstarter idler pinion over the shaft adjacent to the mainshaft because this cannot be replaced when the main body of the clutch is in position. It has a bellville washer and a plain washer below it, and is retained by a circlip.

3   Slide the spacer over the gearbox mainshaft and then lower the clutch outer drum and integral primary drive gear into position so that it meshes with both the primary drive pinion and the kickstarter idler.

4   Position a further plain washer over the mainshaft which protrudes through the clutch outer drum, then position the clutch centre to engage with the splines on the end of the mainshaft. Replace the domed spring washer and the centre nut retaining the clutch centre in position (right-hand thread).

5   Lock the gearbox in position by wrapping chain around the final drive sprocket and holding both ends in a vice. The centre retaining nut of the clutch assembly can now be tightened. Release the chain from the vice and remove it from the sprocket.

6   The clutch plates can now be inserted one at a time, commencing with a plain plate. Note that a rubber ring is interposed between each following set of inserted and plain plates, so that the plain plates cannot come into contact with one another when the clutch is withdrawn and create clutch

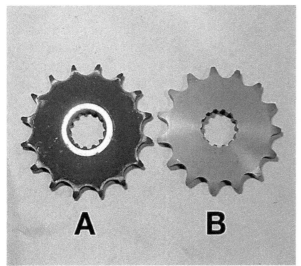

34.1a Comparison of old (A) and new (B) sprockets

34.2 Sprocket fits over splined shaft

34.3 Do not omit tab washer

35.1 Domed washer fits below pinion retaining nut

36.2 End of return spring locates with crankcase peg

37.1 Short rod locates with slot in drive pinion

37.2 Drive pinion is retained by a washer and circlip

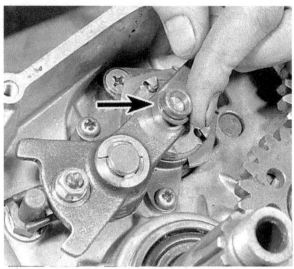

38.3a Do not omit shouldered collar, easy to mislay

38.3b Slot in gear change shaft end slides over shouldered collar

38.4 Circlip retains gear change lever shaft

39.1 Fit spacing washer first

39.2a Kickstarter idler pinion is preceded by belville washer

39.2b Pinion must be fitted before clutch is built up

39.3a Spacer fits over mainshaft before ...

39.3b ... Clutch outer drum can be lowered into position

39.4 Domed washer and nut retain clutch centre

39.6a Rubber rings are interposed between plain clutch plates

39.6b Do not omit clutch 'mushroom' before pressure plate is fitted

40.2 Replace the caged needle roller small end bearings

40.3 Warm pistons if gudgeon pins are tight fit

40.4 Piston rings have expanders fitted behind them

41.2 Piston ring clamps aid fitting of pistons

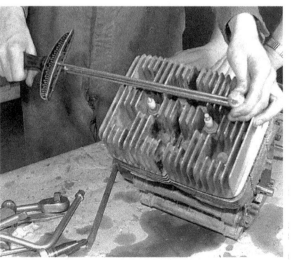

42.3 Use torque wrench to tighten down to correct setting

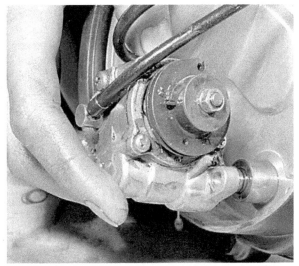

43.1 Oil pump is retained by two screws, after removing drive

44.1a Woodruff key retains rotor on crankshaft taper

44.1b Centre bolt passes through contact breaker cam

44.2a Dowel ensures correct location of stator cover

44.2b Grommet seals cable lead-through

44.3 Do not forget to connect neutral switch

45.1 Lift engine into frame from right-hand side

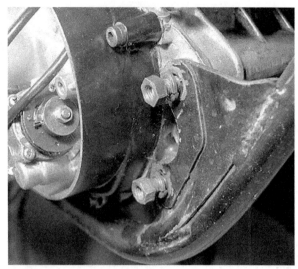

45.2a Fit front engine bolts first

45.2b Also rear bolt near swinging arm

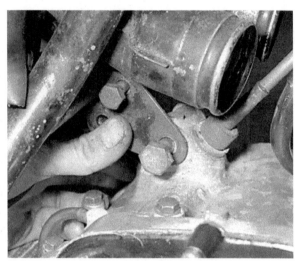

45.3 Upper rear bolts pass through small engine plates

46.1a Thread main oil feed pipe through grommet in outer cover

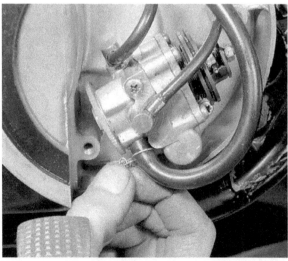

46.1b Reconnect with oil pump and locate spring clip

noise. The sequence of assembly is therefore plain plate, rubber ring, inserted plate, plain plate, rubber ring etc. Before the pressure plate is positioned, make sure the clutch push rod 'mushroom' has been greased and inserted into the hollow end of the mainshaft.

7 Replace the six clutch springs in the clutch pressure plate and tighten down the retaining screws until the full depth of thread is engaged.

40 Engine reassembly: replacing the pistons and piston rings

1 Position the engine to rest on the base of the crankcase. Pad the mouth of each crankcase with clean rag prior to fitting the pistons and piston rings, so that any displaced parts will be prevented from falling in.
2 Replace the caged needle roller bearings in each small end, then fit the pistons and gudgeon pins, checking to ensure that they are replaced in their original locations. Note that the pistons have an arrow stamped on the crown, which must face forwards.
3 If the gudgeon pins are a tight fit in the piston bosses, the pistons can be warmed with warm water to effect the necessary temporary expansion. Oil the gudgeon pins and piston bosses before the gudgeon pins are inserted, then fit the circlips, making sure that they are engaged fully with their retaining grooves. A good fit is essential, since a displaced circlip will cause extensive engine damage. Always fit new circlips, NEVER re-use the old ones.
4 Check that the piston rings are fitted correctly, with their ends either side of the ring pegs and the expanders behind them. If this precaution is not observed, the rings will be broken during assembly.

41 Engine reassembly: replacing the cylinder barrels

1 Place a new cylinder base gasket over the retaining studs and lubricate each cylinder bore with clean engine oil. Arrange the left-hand piston so that it is at top dead centre (TDC) and lower the left-hand cylinder down the retaining studs until contact is made with the piston. The rings can now be squeezed one at a time until the cylinder barrel will slide over them, checking to ensure that the ends are still each side of the ring peg. Great care is necessary during this operation, since the rings are brittle and very easily broken.
2 Although the cylinder barrels have a good lead-in, to facilitate entry of the piston rings, a piston ring clamp can be used as an alternative to the hand feed method. Here again, care must be taken to ensure that the rings are correctly positioned in relation to the piston ring pegs.
3 When the rings have engaged fully with the cylinder bore, withdraw the rag packing from the crankcase mouth and slide the cylinder barrel down the retaining studs, so that it seats on the new base gasket (no gasket cement).
4 Repeat this procedure for the right-hand cylinder, after arranging the right-hand piston so that it is at TDC.

42 Engine reassembly: replacing the cylinder heads

1 Because the cylinder heads have a full hemisphere and have central spark plugs, they are identical and fully interchangeable.
2 Place a new copper cylinder head gasket on the top of each cylinder barrel, using a smear of grease to retain them in position. Fit each cylinder head in turn, taking care that the cylinder head gasket is not displaced or distorted during the initial tightening down.
3 Each cylinder head has four sleeve bolts, which must be tightened evenly, in a diagonal sequence. This is most important, since distortion will occur if this precaution is not observed. Use a torque wrench to achieve the final setting of 15 ft lb.
4 Replace the spark plugs in order to prevent any extraneous

material from dropping into the engine, especially whilst it is being refitted into the frame.

43 Engine reassembly: replacing the right-hand crankcase cover and oil pump

1 The oil pump is attached to the right hand crankcase cover and need not be disturbed unless the cover is broken, necessitating renewal. If it is necessary to transfer the pump from one case to another, follow the procedure given in Chapter 2, Section 13.
2 Since the outer crankcase cover contains the gearbox oil filler and has to retain the oil content, a good joint between the outer crankcase and the cover is essential. A gasket is used at the jointing face; provided the jointing surfaces are in good condition and undistorted, the joint can be made dry. If there is any doubt, give both surfaces a light coating of gasket cement.
3 The cover is retained by eight cross head screws. Care is necessary to ensure the oil pump driving pinion meshes with the primary drive pinion; if the cover is forced into position teeth may be broken from the plastic drive pinion. The cover has two dowels to aid location.

44 Engine reassembly: replacing the alternator

1 Provided care is exercised when replacing the engine unit in the frame, there is no reason why the alternator should not be fitted at this stage. Replace the woodruff key in the tapered left-hand end of the crankshaft and position the rotor on the shaft so that the keyways align. The rotor is retained by a centre bolt passed through the centre of the ignition cam. The latter is pegged, to ensure it is located correctly.
2 Tighten the centre bolt, then replace the contact breaker assembly and brush gear cover, which is attached by three long cross head screws. The wiring harness feeds through an aperture in the rear of the upper crankcase casting and is sealed by a rubber grommet. A clip, through which one of the upper crankcase bolts passes, acts as a cable retainer. A small dowel ensures that the cover can be replaced in one position only.
3 Reconnect the short lead to the neutral indicator switch by the spade end connector and join up the main harness with the snap connector provided. A small, square-shaped rubber grommet seals the gap in one of the crankcase ribs, through which the wire passes. The engine unit is now ready for refitting in the frame.

45 Refitting the engine/gear unit in the frame

1 Place the machine on the centre stand, so that it is standing rigidly on firm ground. Lift the engine unit and with the aid of an assistant, slide the engine unit into the frame from the right-hand side. The front engine mountings are waisted, to correspond with the design of the front of the crankcase, so that there is sufficient clearance for insertion.
2 When the engine unit is in approximately the correct position, fit the two front engine bolts, and the engine bolt close to swinging arm pivot. Do not tighten the bolts at this stage.
3 Replace the short upper rear engine bolt which passes through a lug cast into the crankcase and two small engine plates. Replace the two bolts that retain these upper engine plates, then tighten all the bolts to the recommended torque settings of 15 lf lb for the 8 mm bolts and 25 ft lb for the 10 mm bolts.

46 Engine reassembly: reconnecting the oil pipes and carburettors

1 Reconnect the main oil feed pipe to the oil pump inlet after threading the pipe through the rubber grommet at the top of the right-hand crankcase cover. The pipe is held in position by a

small wire clip.

2 Reconnect the flexible oil pipes to the inlet ports, using new fibre washers for the banjo union joints. Reconnect the tachometer drive.

3 Fit the twin carburettors to the flexible induction stubs. Note that the two carburettors are inter-connected by a short rubber balance pipe, and that the carburettor with the choke lever is fitted on the left. Each carburettor is retained by a metal clip around the flexible rubber intake into which the carburettor body pushes.

4 Replace the carburettor tops, complete with the throttle slide and needle assembly. The slot in the side of each slide must locate with a projection within the mixing chamber, before the slide can be lowered into its correct position and the top screwed home. Before reconnecting the carburettor intakes with the air cleaner assembly, check that the slides are synchronised correctly, as described in Chapter 2, Section 8. A short rubber hose connects each intake with the air cleaner; it is a push on fit over the air cleaner joint and is secured to the carburettor intake by means of a metal clamp.

5 Group the carburettor drain pipes without sharp bends so that they will vent freely.

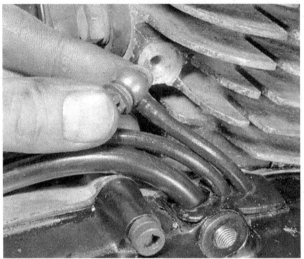
46.2a Use new washers at inlet port connections

47 Engine reassembly: reconnecting, bleeding and setting the oil pump

1 Reconnect the wire control cable linking the oil pump with the twist grip throttle. It is best to arrange the wire in a loop and engage the nipple with the oil pump pulley, before the cable is seated in the pulley groove.

2 Because the oil feed pipe now contains air, it is necessary to bleed the oil pump until all the air bubbles are removed. Check that the oil tank is filled with oil, then unscrew and remove the small cross head screw, which has a fibre washer beneath the head, from the oil pump body. Rotate the plastic wheel with the milled edge at the rear of the oil pump, in a CLOCKWISE direction (as denoted by the arrow marking) and continue turning until the oil commences to flow from the outlet which the bleed screw normally seals. Continue turning until all air bubbles have been eliminated from the main feed, then replace the screw and washer.

3 To check whether the pump opening is correct, follow the procedure given in Chapter 2, Section 15. If the pump was set up correctly initially, it is improbable that significant changes in setting will be required. DO NOT OMIT THIS CHECK UNDER THE ASSUMPTION IT MUST BE CORRECT.

4 Replace the semi-circular cover over the oil pump, which is retained by three cross head screws.

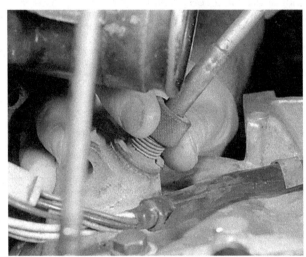
46.2b Reconnect the tachometer drive cable

48 Engine reassembly: replacing the left-hand crankcase cover

1 Before replacing the left-hand crankcase cover, check that the contact breaker gap and ignition timing is correct for both cylinders. An aperture in the generator cover has a pointer which will align with timing marks on the rotor makes checking of the ignition timing quite simple. See Chapter 3, Section 9 for details. Loop the final drive chain over the gearbox sprocket.

2 Replace the clutch cable and small portion of crankcase lip which acts as a cable stop in the top of the outer crankcase housing. Reconnect the cable nipple with the clutch release mechanism in the inside of the left-hand crankcase cover and refit the cover; this is retained by three cross head screws. To make cable fitting easy, slacken off the adjuster at the handlebar lever, or even remove the cable from the lever itself. Before the cover is replaced, do not omit to replace the clutch push rod, which is preceded by a ball bearing, and the plastic cover over the gear change lever shaft. The push rod should be well greased, prior to insertion. The ball bearing at the other (left-hand) end is best retained within the clutch operating mechanism of the left-hand crankcase cover by a dab of grease.

46.4 Check that slides locate correctly before tightening tops

47.1a Thread control cable through cover and connect adjuster

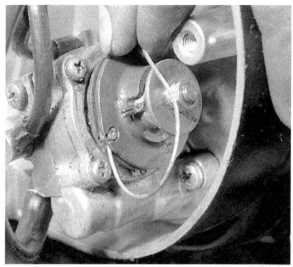

47.1b Engage lower end of cable with pump pulley

47.4 Oil pump cover is retained by three screws (sockets shown are non-standard)

48.1 Aperture in stator cover indicates timing accuracy

48.2a Refit clutch cable stop before outer cover is positioned

48.2b Cable engages with actuating mechanism in rear of cover

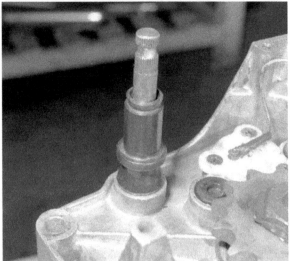
48.2c Do not omit cover over gear change lever shaft

48.3 Alternator cover is retained by three long screws

48.5 Clutch adjustment cover has dowel location

49.3 New exhaust pipe gaskets are essential

50.2a Kickstarter is retained on splines by a pinch bolt ...

50.2b ... in similar fashion to that of gear change lever

3 Refit the circular generator cover, and secure with the three long cross head screws which pass through the left-hand outer cover.

4 There is no gasket at the left-hand crankcase cover joint.

5 Before fitting the small insert in the left-hand cover which masks off the clutch operating mechanism, check that clutch adjustment is correct. There should be a small amount of play when the adjusting screw is turned in a clockwise direction, before pressure is applied to the clutch push rod. If necessary, fit the clutch cable to the handlebar lever and after taking the slack out of the cable with the handlebar lever adjuster, adjust the operating mechanism so that there is about 3/32" play at the handlebar lever before the clutch action commences. Adjustment is effected by slackening the locknut of the adjusting screw and turning the screw clockwise to decrease the clearance or anti-clockwise to increase it, before tightening the locknut and re-checking.

6 When replacing the clutch adjustment cover, which is retained by two cross head screws, note that it can be replaced in one position only due to the dowel method of location employed.

7 Reconnect the final drive chain. It is easiest to fit the spring link when both ends of the chain are pressed into the rear wheel sprocket.

49 Engine reassembly: replacing the exhaust pipes and silencers

1 Replace the exhaust pipes and silencers, which are constructed on the one-piece principle. Feed the exhaust pipes through the gap between the footrests and the crankcase covers, then rotate the whole assembly through 90° so that the exhaust pipes are approximately in line with the exhaust ports.

2 Fit the silencer retaining bolts first. The bolts pass through the lower part of the plate that holds the pillion footrests; each silencer flange has a rubber bush insert, which should be renewed if at all faulty.

3 When attaching the exhaust pipe flanges to the cylinder barrels, always use new gaskets of the correct type. Pull the flanges down evenly, so that the exhaust pipe joint is a good airtight fit.

50 Engine reassembly: completion and final adjustments

1 Remake the electrical connections from the alternator harness. The various connectors should be protected by the plastic sleeve which surrounds them, close to the right-hand carburettor.

A clip attached to one of the top, left-hand crankcase bolts acts as a guide for the wiring harness, and prevents the harness from chafing or rubbing. Reconnect the battery and the spark plug leads, and check that the lights and ignition circuit are live when switched on.

2 Replace the kickstarter and the gear change lever. Both fit on splines and are retained by a pinch bolt. Check that they are at the desired operating angle before tightening the pinch bolt.

3 Replace the petrol tank, making sure that the mounting rubbers engage with the retainers close to the steering head. Connect the pipe uniting both halves of the petrol tank, on the underside, and make sure it is retained by wire clips. Connect the feed pipes to the carburettors and turn the petrol tap to the off position. Refill the tank with petrol.

4 Check the contents of the oil tank and top up if necessary with oil SAE 30 two-stroke grade. Remove the filler cap from the right-hand crankcase cover and add 1500 cc of IOW/30 oil. Check with the dipstick that the level is correct.

51 Starting and running the rebuilt engine

1 When the initial start-up is made, run the engine slowly for the first few minutes, especially if the engine has been rebored or a new crankshaft fitted. Check that all the controls function correctly and that there are no oil leaks before taking the machine on the road. The exhausts will emit a high proportion of white smoke during the first few miles, as the excess oil used whilst the engine was reassembled is burnt away. The volume of smoke should gradually diminish until only the customary light blue haze is observed during normal running. It is wise to carry a spare pair of spark plugs during the first run, since the existing plugs may oil up due to the temporary excess of oil.

2 Remember that a good seal between the pistons and the cylinder barrels is essential for the correct functioning of the engine. A rebored two-stroke engine will require more carefully running-in, over a longer period, than its four-stroke counterpart. There is a far greater risk of engine seizure during the first hundred miles if the engine is permitted to work hard.

3 Do not tamper with the exhaust system or run the engine without the baffles fitted to the silencer. Unwarranted changes in the exhaust system will have a very marked effect on engine performance, invariably for the worst. The same advice to dispensing with the air cleaner or the air cleaner element.

4 Do not on any account add oil to the petrol under the mistaken belief that a little extra oil will improve the engine lubrication. Apart from creating excess smoke, the addition of oil will make the mixture much weaker, with the consequent risk of overheating and engine seizure. The oil pump alone should provide full engine lubrication.

52 Fault diagnosis - Engine

Symptom	Reason/s	Remedy
Engine will not start	Defective spark plugs	Remove plugs and lay on cylinder head. Check whether spark occurs when engine is kicked over.
	Dirty or closed contact breaker points	Check condition of points and whether gap is correct.
	Discharged battery	Check whether lights work. If battery is flat, remove and charge.
	Air leak at crankcase or worn crankshaft oil seals	Flood carburettors and check whether petrol is reaching the plugs.
Engine runs unevenly	Ignition and/or fuel system fault	Check as though engine will not start.
	Blowing cylinder head gasket	Oil leak should provide evidence. Replace gasket
	Incorrect ignition timing	Check and if necessary adjust.
	Carburettors out of balance	Refer to Chapter 2 and adjust.
	Choked silencers	Remove baffles and clean.
Lack of power	Incorrect ignition timing	See above.
	Fault in fuel system	Check system and vent in filler cap.
	Choked silencers	See above.
White smoke from exhaust	Too much oil	Check oil pump setting.
	Engine needs rebore	Rebore and fit oversize pistons
	Tank contains two-stroke petroil and not straight petrol	Drain and refill with straight petrol.
Engine overheats	Pre-ignition and/or weak mixture	Check carburettor settings, also grade of plugs fitted.
	Lubrication failure	Stop engine and check oil pump setting. Is oil tank dry?

53 Fault diagnosis - Gearbox

Symptom	Reason/s	Remedy
Difficulty in engaging gears	Selector forks or rods bent	Replace
	Broken springs in gear selector mechanism	Check and replace
	Clutch drag	See next Section
Machine jumps out of gear	Worn dogs on ends of gear pinions	Strip gearbox and replace worn parts.
	Sticking camplate plunger	Remove plunger cap and free plunger assembly.
Kickstarter does not return	Broken return spring	Remove right hand crankcase cover and replace spring.
Kickstarter slips or jams	Worn rachet assembly	Remove right hand crankcase cover, dismantle kickstarter assembly and replace worn parts.
Gear change lever does not return to normal position	Broken return spring	Remove right hand crankcase cover and replace spring.

54 Fault diagnosis - Clutch

Symptom	Reason/s	Remedy
Engine speed increases but machine does not respond	Clutch slip	Check whether clutch adjustment still has free play. Check thickness of linings and replace if near wear limit.
Difficulty in engaging gears, gear changes jerky and machine creeps forward, even when clutch is withdrawn fully	Clutch drag	Check clutch adjustment to eliminate excess play. Check whether clutch centre and outer drum have indented slots.
	Clutch assembly loose on mainshaft	Check tightness of retaining nut.
Operating action stiff	Bent push rod	Replace.
	Dry push rod	Lubricate.
	Damaged, trapped or frayed control cable	Check cable and replace if necessary. Make sure cable is lubricated and has no sharp bends.

Chapter 2 Fuel system and lubrication

Contents

Specifications

Petrol tank

Capacity	...	...	...	...	...	...	...	3.2 Imp gallons	12 litres

Oil tank

Capacity	...	...	...	...	...	...	...	3½ Imp pints	2 litres

Carburettors

									YDS7	YR5
Make	...	...	...	...	...	...	...	...	Mikuni	Mikuni
Type	...	...	...	...	...	...	...	...	VM26SC	VM28SC
Main jet	...	...	...	...	...	...	...	...	100	110
Jet needle	...	...	...	...	...	...	...	...	5DP7	5DP7
Needle jet	...	...	...	...	...	...	...	...	0–0	0–0
Throttle valve cutaway	...	...	...	...	...	...		2	2
Pilot jet	...	...	...	...	...	...	...	...	40	40
Air Screw	...	...	...	...	...	...	...	...	1½ – 1¾ turns	1½ – 1¾ turns
Starter jet	...	...	...	...	...	...	...	100*	100*
Float valve seat	...	...	...	...	...	...		2.0	2.0

*left-hand carburettor only

Oil pump

Minimum stroke tolerance		...	...	...	...	...	0.20 – 0.25 mm (0.008" – 0.010")	0.20 – 0.25 mm (0.008 – 0.010")
Oil viscosity	...	...	...	...	...	...	...	SAE 30	SAE 30

1 General description

The fuel system comprises a petrol tank from which petrol is fed by gravity, via a petrol tap with a built-in bowl-type filter, to the float chambers of the twin Mikuni carburettors.

For cold starting, the left-hand carburettor is fitted with a hand operated choke. This provides the rich mixture necessary for a cold start and can be opened as soon as the engine will accept full air under normal running conditions.

Unlike many two-strokes, the Yamaha 250/350 cc twins do not require a petrol/oil mix for lubrication. Oil for lubricating the engine is contained within a separate, side-mounted oil tank, from which it is fed to an engine-driven oil pump, attached to the right-hand side of the crankcase. Oil from the pump is delivered to a drilling in the inlet passage of each cylinder barrel and is drawn into the engine with the incoming mixture. The two-stroke engine depends on the compression of the incoming mixture before it is transferred to the combustion chamber via the transfer ports. Thorough lubrication of the bottom end of the engine as well as the cylinders and pistons is therefore essential. An added refinement is a direct link between the oil pump and the throttle so that the oil pump opening varies according to engine demand.

2 Petrol tank: removal and replacement

1 Although it is not necessary to remove the petrol tank when the engine unit is removed from the frame, better access is gained and there is less risk of damage to the painted surface if the tank is out of the way. Apart from occasions such as these, there is rarely any need to remove the tank unless rust has formed inside as the result of long storage or if it needs repainting.

2 The petrol tank is not rigidly attached to the frame in any

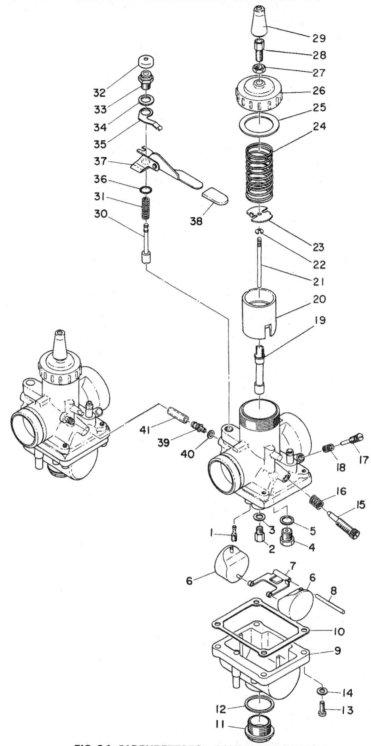

FIG. 2.1. CARBURETTORS — COMPONENT PARTS

1 Pilot jet	10 Float chamber gasket	20 Throttle valve	31 Starter plunger spring
2 Main jet	11 Drain plug	21 Needle	32 Plunger cap cover
3 Washer for main jet	12 Washer for drain plug	22 Needle clip	33 Plunger cap
4 Float needle seating	13 Screw	23 Throttle spring seating	34 Starter lever washer
5 Washer for float needle	14 Spring washer	24 Throttle spring	35 Starter lever plate
seating	15 Throttle stop screw	25 Packing	36 'O' ring
6 Float	16 Spring for pilot jet	26 Mixing chamber top	37 Starter lever - left-hand
7 Float arm	adjusting screw	27 Cable adjuster lock nut	38 Cap
8 Float hinge pin	17 Pilot jet adjusting screw	28 Cable adjuster	39 Nipple
9 Float chamber body	18 Spring for throttle stop screw	29 Rubber cap	40 Washer
	19 Needle jet	30 Starter plunger (choke)	41 Pipe joint

FIG. 2.2. OIL PUMP – COMPONENT PARTS

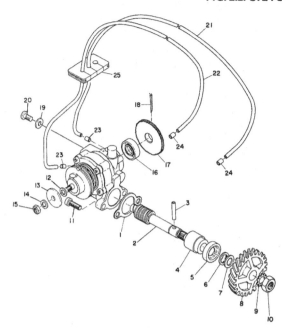

1   Pump case gasket
2   Worm shaft
3   Dowel pin
4   Worm shaft outer bush
5   Oil seal
6   Circlip
7   Washer
8   Pump drive gear pinion
9   Toothed washer
10 Nut
11 Screw - 2 off
12 Plunger shim
13 Adjusting plate
14 Spring washer
15 Nut
16 Oil seal
17 Starter plate
18 Split pin
19 Bleed screw
20 Bleed screw
21 Left-hand delivery pipe
22 Right-hand delivery pipe
23 Delivery pipe clip - 2 off
24 Delivery pipe clip - 2 off
25 Oil pipe retainer

way. It is retained by rubber buffers at the front and rear, which are compressed when the tank is pushed into position over the top frame tube. The forward mounted rubbers engage with retaining cups attached to each side of the gusset at the rear of the steering head. The nose of the dowel seat presses lightly on the rear of the tank.

3   Before the petrol tank can be removed, it is necessary to detach the twin petrol pipes from the carburettor float chambers. Each is retained by a small wire clip. Detach the pipe that joins both halves of the petrol tank, on the underside. This too is retained at each end by a wire clip. The tank must be drained before it can be lifted from the frame, either by the petrol tap or by detaching one end of the underside cross pipe. The dual seat must be raised to free the rear end of the tank.

4   When replacing the tank, check that the mounting rubbers are in good condition and are located correctly. Do not forget to replace and secure the pipe joining both halves of the petrol tank, before the tank is refilled.

3   Petrol tap: removal, dismantling and replacement

1   The petrol tap is secured to the left-hand underside of the petrol tank by two concealed cross head screws, which are exposed when the filter bowl is removed. The filter bowl threads into the main body of the petrol tap and can be removed by applying a spanner to the hexagon in the base of the bowl. There is a sealing washer between the filter bowl and the main body of the petrol tap to preserve a petrol-tight joint; the filter gauze is located above the filter bowl, attached to the main body of the petrol tap by a small cross screw.

2   There is seldom need to disturb the main body of the petrol tap. In the event of a leak at the operating lever, the complete lever assembly can be dismantled (provided the petrol tank is drained first) with the main body undisturbed. Remove the two cross head screws that retain the tap lever plate in position and withdraw the lever complete with plate, crinkle washer and the valve insert behind the lever. The valve is the item to be replaced if leakage occurs; it is composed of a synthetic rubber material which may commence to disintegrate after an extended period of service.

3   The main body of the tap is secured to a flange on the underside of the petrol tank by two cross-head screws within the tap body. There is a gasket between the flange of the petrol tap body and the underside of the petrol tank, to preserve a petrol-tight joint. This gasket should be renewed, irrespective of its condition, whenever the tap body is removed and replaced.

4   Before reassembling the petrol tap, check that all parts are clean, especially the two tubes (short tube reserve, long tube main feed) which extend into the petrol tank, the filter and filter bowl assembly. A new gasket should be fitted to the filter bowl assembly in order to effect a satisfactory seal.

5   Do not over-tighten any of the petrol tap components during reassembly. The castings are in a zinc-based alloy, which will fracture easily if over-stressed. Most leakages occur as the result of defective seals.

4   Petrol feed pipes: examination

1   The plastic pipes used for the various connections between the petrol tank and the carburettors are a thin-walled push-on type, secured by wire clips. Renewal is seldom required, unless it becomes hard or splits. Always replace the pipes in the same order because the ends sometimes take a permanent form of the fitting over which they are connected.

5   Carburettors: removal

1   Before removing the carburettors it is necessary to detach the petrol feed pipe at the point where it joins each float chamber. Pinch together the two 'ears' of the clip and slide the clip up the pipe so that the end of the pipe is free to be pulled from the float chamber connection

2   Each carburettor is connected to the air cleaner by means of a short length of hose secured by a clip around each intake. Slacken the cross head screw that retains each clip and pull the hoses away. The other end of each hose is a push fit over the inlet stub to the air cleaner housing.

3   Unscrew the top of each mixing chamber and remove the top complete with the cable, throttle valve and needle assembly. It is a wise precaution to tape these parts to a near-by frame tube so that they are not damaged as dismantling continues.

4   Slacken the cross head screw in the clamp retaining each carburettor in its flexible rubber mounting stud. The carburettors can now be pulled away from their mountings and placed on the workbench for further dismantling. Note that they are interconnected on the inside faces by a short rubber 'balance' pipe which aids their synchronisation.

2.1 When drained and pipes disconnected, tank can be lifted off

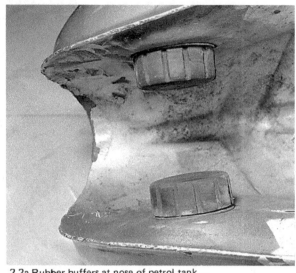

2.2a Rubber buffers at nose of petrol tank ...

2.2b ... retaining cups welded to steering head

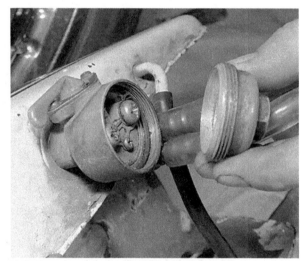

3.1a Filter bowl unscrews from base of petrol tap

3.1b Crescent-shaped filter is retained by a screw

3.2 Tap lever assembly is held by plate and two screws

3.3 Two screws within tap body retain tap to tank

5.2 Metal clip retains carburettor to flexible mounting

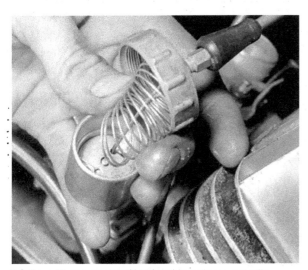

5.3 Top will unscrew; has throttle valve, spring and needle attached

6 Carburettors: dismantling and examination

1   To separate the float chamber, invert the carburettor and remove the four small screws and spring washers holding the float chamber in position. There is a gasket between the float chamber body and the mixing chamber, to maintain a petrol-tight joint. The old gasket should be discarded.

2   The twin plastic floats, will lift out of the float chamber. They fit over the two metal rods projecting vertically from the base of the float chamber and have the dished portion uppermost, to clear the float needle arm. They are separate units.

3   The float arm is attached to the main body of the carburettor and is displaced when the pivot pin is removed. Persistent flooding is invariably caused by a leaking float, which will cause the petrol level to rise, by dirt on the float needle, or its seating. Renewal of the defective float is the only remedy in the case of a leaking float; it is not possible to effect a practicable repair. Dirt on the float needle or its seating is best removed with a jet of compressed air.

4   A large diameter plug in the base of each float chamber allows access to the main jet of the carburettor without need to remove the float chamber. It is important that the fibre washer that seals the plug is in good condition, otherwise a continual leakage will occur.

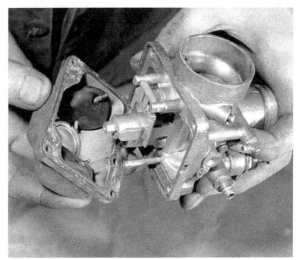

6.1 Invert carburettor to remove float chamber

5   The needle jet screws into the mixing chamber from the top and is located immediately above the main jet. This jet is subject to wear and should be renewed if the petrol consumption appears unduly high.

6   The throttle valve is still attached to the top of the mixing chamber by means of the throttle cable and return spring. To release the throttle valve, lift the return spring and remove the metal seating that fits over the needle. This serves the dual function of providing a seating for the spring and by a bent tab which forms a locking device to prevent the throttle cable from being detached. When the seating is removed, the throttle cable nipple can be slipped out of the throttle valve and the throttle valve detached. The needle will lift out, with the retaining clip which holds it in the correct notch. When the needle jet is renewed, the needle should be renewed too since they work in conjunction with one another.

7   Before reassembling the carburettor in the reverse order of that given for dismantling, make sure all the component parts are clean. Check that the needle is not bent, by rolling it on a sheet of plain glass. examine the throttle slide; signs of wear will be evident on the polished outer surface.

8   Never use wire or any pointed instrument to clear a blocked jet or any of the internal air passages in the mixing chamber body. It is only too easy to enlarge the small precision drilled

orifices and cause carburation changes which will prove very
difficult to rectify. Always use compressed air; even a jet of
air from a tyre pump should suffice.

9 When replacing the throttle valves, make sure the slot in the
base of each valve registers with the projection inside the mixing
chamber, so that the valve will seat correctly. It is also important
to check that the needle suspended from each throttle valve has
entered the needle valve, otherwise there is risk of damaging
both the needle and the jet.

10 Similar advice to that given about the petrol tap applies to
the carburettors, which are made of the same die-cast alloy.
They will fracture or distort badly if force is used during
reassembly.

7 Carburettors: checking the settings

1 The various jet sizes, throttle valve cutaway and needle
position are predetermined by the manufacturer and should not
require modification. Check with the Specifications list at the
beginning of this Chapter if there is any doubt about the valves
fitted.

2 Slow running is controlled by a combination of the throttle
stop and pilot jet settings. Adjustment should be carried out as
explained in the following Section. Remember that the
characteristics of the two-stroke engine are such that it is
extremely difficult to obtain a slow, reliable tick-over at low
rpm. If desired, there is no objection to arranging the throttle
stop so that the engine will shut off completely when the
throttle is closed. Unlike a petroil-lubricated engine, the oil used
for engine lubrication is injected into the inlet passage of each
cylinder barrel, behind the closed throttle slide. In consequence
there is no risk of the engine 'drying up' when the machine is
coasted down a long incline, if the throttle is closed.

3 As a rough guide, up to 1/8th throttle is controlled by the
pilot jet, 1/8th to 1/4 by the throttle valve cutaway, 1/4
to 3/4 throttle by the needle position and from 3/4 to full
by the size of the main jet. These are only approximate
divisions, which are by no means clear cut. There is a certain
amount of overlap between the various stages.

4 The normal setting of the pilot jet screw is approximately
one and three quarter complete turns out from the closed
position. If the engine 'dies' at low speed, suspect a blocked pilot
jet in the carburettor of the cylinder concerned.

5 Guard against the possibility of incorrect carburettor adjustments which will result in a weak mixture. Two-stroke engines are
very susceptible to this type of fault, causing rapid overheating
and often subsequent engine seizure. Changes in carburation
leading to a weak mixture will occur if the air cleaner is
removed or disconnected, or if the silencers are tampered
with in any way. Above all, do not add oil to the petrol,
in the mistaken belief that it will aid lubrication. The
extra oil will only reduce the petrol content by the ratio of oil
added, and therefore cause the engine to run with a permanently weakened mixture.

8 Synchronising twin carburettors

1 For optimum performance and even running, it is important
that both carburettors are in phase, so that they open and close
together and have the same settings throughout the entire throttle
range. If this type of check is not carried out at regular intervals,
one cylinder may do all the work, whilst 'carrying' the other. The
usual symptoms of poor balance between the two carburettors
are difficulty in starting and uneven running at low speeds,
accompanied by a noticeable lag when accelerating.

2 To check whether the carburettors are correctly synchronised, temporarily disconnect the intake hoses between the mouth
of each carburettor and the air cleaner. Unscrew both throttle
stop screws (anticlockwise) until both throttle valves close
completely. Turn the twistgrip slowly and check whether both

6.2 Float chamber has two plastic floats, not connected

6.3a The float arm is attached to the main body of the carburettor

6.3b The float needle is located beneath the float arm

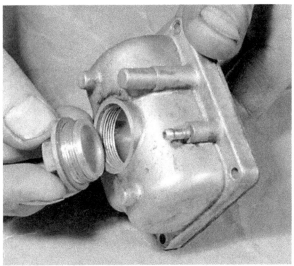

6.4a Large plug in base gives access to main jet

6.4b Main jet screws into jet holder

6.6 Cable detaches from throttle valve after removing metal seating

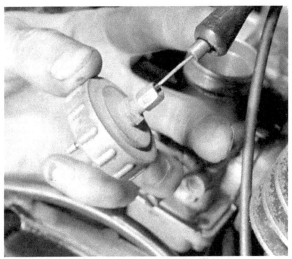

8.2 Adjuster in each carburettor top facilitates cable adjustment

9.1 Air cleaner element is of corrugated paper type

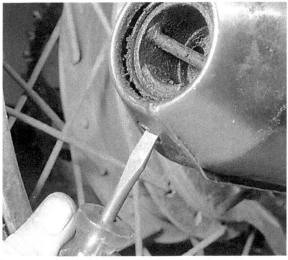

11.3a Withdraw screw to remove silencer baffles

throttle valves lift at exactly the same time. If they do not rise in unison, one cable has more slack than the other. This should be taken out by unscrewing the adjuster at the top of the carburettor which is lagging, in an anti-clockwise direction until both throttle valves lift simultaneously.

3   Continue opening the throttle by means of the twist grip and check that both throttle valves clear the carburettor intakes simultaneously. This is a cross-check that the cable adjustments are correct; if the throttle valves are not exactly in step, close the throttle and start all over again. Most probably one valve is not seating correctly or the cable adjustment is still incorrect. When the two throttle valves are exactly in step throughout the range, re-connect the air cleaner hoses.

4   Remove the plug cap from the left-hand spark plug and start the engine. Screw in (clockwise) the throttle stop screw of the right-hand carburettor until the engine will continue to run on one cylinder at a reasonably low speed. Check whether adjustment of the pilot jet screw from the recommended setting of one and three quarters of a complete turn out has any noticeable effect on even running and readjust the throttle stop screw as necessary. Stop the engine.

5   Repeat the above operation with the right-hand spark plug cap removed and adjust the left-hand carburettor. Stop the engine when the adjustment seems correct.

6   Re-start the engine with both plug caps connected. Most probably the tick-over will now be too fast, in which case the throttle stop screws of each carburettor can now be screwed out an equal amount (a little at a time) until the desired engine speed is attained. Normally this is in the region of 1,300 to 1,400 rpm, as indicated by the tachometer.

11.3b Baffles will pull out as a complete unit

9   Air cleaner: removing, cleaning and replacing the element

1   The air cleaner is located beneath the nose of the dual seat, which must be raised to gain access to the element. The lid of the air cleaner box is retained by a rubber strap, which must be detached before the lid can be removed. The element, (corrugated paper type) will lift out.

2   The element is cleaned by blowing it with compressed air or by lightly tapping it so that the loose dust on the surface will be displaced. Care is necessary when handling it because corrugated paper is impregnated with resin and is therefore easily damaged.

3   If the element is damp or contaminated in any way, it must be discarded and a new one fitted. This applies if it is perforated at any point.

4   Irrespective of its condition, the air cleaner element should be renewed every 3,000 miles.

5   Do not **on any account** run the machine with the element removed or with the air cleaner hoses disconnected. If this precaution is not observed, the engine will run with a permanently weak mixture, which will cause overheating and most probably engine seizure. The carburettors are jetted to compensate for the presence of the air cleaner element and the balance is upset when it is removed or disconnected.

13.3a Drive pinion is retained by a nut and washer

10 Crankcase drain plugs

1   Unlike most two-stroke engines, the lower crankcase is not fitted with drain plugs with which to drain the contents of the two separate crankcase compartments.

2   If overflooding of the carburettors causes difficulty in starting the engine, the excess petrol vapour is best removed by unscrewing both spark plugs, turning off the petrol, and kicking over the engine with the throttle wide open until the engine has been thoroughly vented.

13.3b Note rod that engages with slot in pinion

13.4a Circlip fits beneath drive pinion

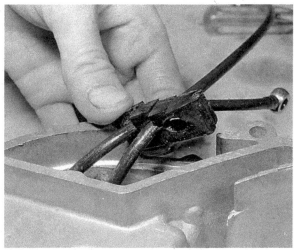

13.4b Oil pipes pass through grommet in outer cover

11 Exhaust pipes and silencers: examining and cleaning the exhaust system

1   Each exhaust pipe and silencer is manufactured as a complete system, and the exhaust pipe cannot be separated from the silencer. This method of manufacture has at least two advantages; air leaks at the exhaust pipe/silencer joint are eliminated and it is less easy to make an unwitting change of silencer, which would otherwise have an adverse effect on both carburation and engine performance.

2   The parts most likely to require attention are the silencer baffles, which will block up with a sludge composed of carbon and oil if not cleaned out at regular intervals. A two-stroke engine is very susceptible to this fault, which is caused by the oily nature of the exhaust gases. As the sludge builds up, back pressure will increase, with a resulting fall-off in performance.

3   There is no necessity to remove the exhaust system in order to gain access to the baffles. They are retained in the end of each silencer by a screw, reached through a hole cut in the underside of each silencer body, close to the end. When the screw is removed, the baffles can be withdrawn.

4   If the build-up of carbon and oil is not too great, a wash with a petrol/paraffin mix will probably suffice as the cleaning medium. Otherwise more drastic action will be necessary, such as the application of a blowlamp flame to burn away the accumulated deposits. Before the baffles are refitted, they must be thoroughly clean, with none of the holes obstructed.

5   When replacing the baffles, make sure the retaining screw is located correctly and tightened fully. If the screw falls out, the baffles will work loose, creating excessive exhaust noise accompanied by a marked fall-off in performance.

6   Do not run the machine without the baffles in the silencer or modify the baffles in any way. Although the changed exhaust note may give the illusion of greater power, the chances are that performance will fall off, accompanied by a noticeable lack of acceleration. There is also risk of prosecution by causing an excessive noise. The carburettors are jetted to take into account the fitting of silencers of a certain design and if this balance is disturbed, the carburation will suffer accordingly.

12 The lubrication system

1   Unlike many two-strokes, the Yamaha 250/350 cc twins have an independent lubrication system for the engine and do not require the mixture of a measured quantity of oil to the petrol content of the fuel tank in order to utilise the so-called 'petroil'

method. Oil of the correct viscosity (SAE 30) is contained in a separate oil tank mounted on the left-hand side of the machine and is fed to a mechanical oil pump on the right-hand side of the engine which is driven from the crankshaft by reduction gear. The pump delivers oil at a predetermined rate, via two flexible plastic tubes, to oilways in the inlet passage of each cylinder barrel. In consequence, the oil is carried into the engine by the incoming charge of petrol vapour, when the inlet port opens.

2   The oil pump is also interconnected to the twist grip throttle, so that when the throttle is opened, the oil pump setting is increased a similar amount. This technique, pioneered by a British two-stroke manufacturer in the early 1930's, ensures that the lubrication requirements of the engine are always directly related to the degree of throttle opening. This facility is arranged by means of a control cable looped around a pulley on the end of the pump; the cable is joined to the throttle cable junction box at the point where the cable splits into two for the operation of each carburettor.

13 Removing and replacing the oil pump

1   There is no necessity to remove the oil pump assembly unless the cover itself is damaged and has to be replaced. Under these circumstances the oil pump must be removed as a complete unit, so that it can be fitted to the new crankcase cover.

2   To remove the right-hand crankcase cover, follow the procedure given in Chapter 1, Section 5 (paragraphs 6-8) and Section 8 (paragraphs 1 and 2). There is no necessity to remove the engine from the frame in order to complete this operation, but the oil delivery pipes must be detached from the cylinder barrels.

3   The oil pump is secured to the cover by two cross head screws, but before these can be removed, the drive pinion must first be detached. This is located on the inside of the outer cover and is held by a nut and spring washer. When both are removed, the plastic pinion can be pulled off the oil pump drive shaft and the short metal rod passing through the shaft removed.

4   Unscrew the two cross head screws from the other side of the crankcase cover, remove the circlip from the drive spindle, and lift away the oil pump, complete with the delivery pipes and the grommet through which they pass at the top of the crankcase cover.

5   Refit the oil pump to the replacement crankcase cover, using a new gasket at the oil pump/crankcase cover joint and a new oil seal behind the drive pinion, Replace and tighten the two cross head mounting screws. The remainder of the reassembly is

13.5 Oil seal fits around oil pump drive spindle

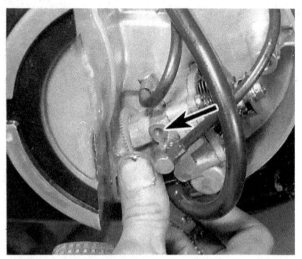

14.2 Oil pump bleed screw location

15.2 Clearance must be correct for effective pump action

15.4 Mark will coincide with pin if setting is correct

accomplished by reversing the dismantling procedure, but **do not** replace the front portion of the crankcase cover because the oil pump must be bled to ensure the oil lines are completely free from air bubbles.

14 Bleeding the oil pump

1   The oil pump must be bled of air if the oil tank to pump pipe has been disturbed or if the oil tank has ever been allowed to run dry.
2   Prior to the bleeding operation check that the pipe connections are well secured by the wire clips and ensure that the tank is topped up to the full line with the recommended oil. Position a container beneath the pump to catch the oil that will be expelled during the bleeding operation. Remove the bleed screw with its sealing washer from the pump and allow the oil to trickle out until the flow is free from air bubbles, then refit the bleed screw and washer; renew the sealing washer if damaged.
3   If the pipes from the pump to the engine have also been disturbed these too must be bled of air. Pull the pump operating cable outer as far out of the casing as possible (so that the pump is in the fully open position) and rotate the white plastic wheel at the side of the pump for several turns.
4   When the bleeding operations have been completed proceed to the next Section to check the pump settings.

15 Checking the oil pump setting

1   Make sure the twist grip is fully closed, then rotate the white plastic pinion used for bleeding the oil pump until the gap between the oil pump pulley at the opposite end of the casing and the body of the oil pump is at its maximum. It will be found that the pulley rises and falls as the pinion is turned, if light pressure is applied with the fingers to the end of the pulley.
2   Check the gap with a feeler gauge. It must be within the range 0.20 - 0.25 mm (0.008" - 0.012") if the pump stroke is correct. To make adjustments, remove the locknut above the plate in the centre of the pulley and lift off the plate. If the clearance was too small, place the appropriate number of 0.1 mm (0.004") shims below the plate before replacing the plate and locknut. If the clearance is too great, remove the appropriate number shims from below the plate. Always re-check after replacing the plate and tightening down the locknut.
3   Adjust the throttle cable so that when the throttle is fully closed there is between 0.5 and 1 mm play. The adjuster is located close to the twist grip.
4   If the pump adjustment is now correct, the mark on the outer face of the oil pump pulley will be directly in line with the guide pin passing through the pulley boss, when the throttle is closed.

5 Check that the pump pulley moves quite freely in each direction as the throttle is opened and closed. Replace the semi-circular end cover.

16 Removing and replacing the oil tank

1 The oil tank is secured to the frame by means of a flexible rubber mounting. The fixings are at the rear of the tank and take the form of a forward mounted bolt which threads into an insert in the tank body and a rear mounted screw that passes through the inner wall of the tank into a recessed area blanked off by the panel containing the capacity plate motif. The screw threads into an insert attached to the back of this panel and helps to hold it in position.

2 Before the tank can be removed, it is necessary to disconnect the breather tube attached to the vent immediately to the rear of the dipstick, the angled synthetic rubber connection between the filler cap and the inlet at the top inner face of the tank retained by a screw drive clip around the rubber hose, and the main feed pipe to the oil pump. When this latter pipe is disconnected by releasing and slipping the wire security clip down the hose and pulling off the connection, the tank can be drained of its oil content.

3 It may be necessary to remove both the battery and the battery carrier to gain access to the screw drive clip around the filler hose, if the clip was positioned badly when the oil tank was fitted originally.

17 Fault diagnosis - Fuel system

Symptom	Reason/s	Remedy
Excessive fuel consumption	Air cleaner choked or restricted	Clean or replace element.
	Fuel leaking from carburettor	Check all unions and gaskets.
	Badly worn or distorted carburettors	Replace.
	Carburettor settings incorrect	Readjust. Check settings with specifications.
Idling speed too high	Throttle stop screws in too far	Adjust screws.
	Carburettor tops loose	Tighten.
Engine sluggish. Does not respond to throttle	Back pressure in silencers	Check baffles and clean if necessary.
Engine dies after running for a short while	Blocked vent hole in filler cap	Clean.
	Dirt or water in carburettors	Remove and clean.
General lack of performance	Weak mixture; float needle sticking in seat	Remove float chambers and check needle seatings.
	Air leak at carburettors or leaking crankcase seals	Check for air leaks or worn seals.

18 Fault diagnosis - Lubrication system

Symptom	Reason/s	Remedy
White smoke from exhausts	Too much oil	Check oil pump setting and reduce if necessary.
Engine runs hot and gets sluggish when warm	Too little oil	Check oil pump setting and increase if necessary.
Engine runs unevenly, not particularly responsive to throttle openings	Intermittent oil supply	Bleed oil pump to displace air in feed pipes.
Engine dries up and seizes	Complete lubrication failure	Check for blockages in feed pipes, also whether oil pump drive has sheared.

Note:
Lubrication failures will occur if a change is made from mineral oil to vegetable-base oils of the 'R' type (or vice-versa) if the engine is not stripped completely and all traces of the original oil removed. Mineral and vegetable oils do not mix but under the action of heat form a rubber-like sludge that will quickly block the internal oilways.

Chapter 3 Ignition system

Contents

Specifications

Alternator

Make	Mitsubishi
Model	AZ201ON
Output	14 volts, 238 watts
Brushes	two
Size of brushes	6 x 7 x 11 mm
Wear limit	6 mm
Brush spring free length	9 mm
Brush spring tension	620 g
Stator coil resistance	4.0 – 4.5 ohms
Rotor coil resistance	0.3 – 0.35 ohms

Regulator unit

Make	Mitsubishi
Model	RN 2226 – J3

Ignition coils

Make	Hitachi
Model	CM11 – 50

Spark plugs

Make	NGK
Type	B – 8HCS (YDS – 7) B – 9HC (YR5)
Reach	½ inch
Gap	0.5 - 0.7 mm (0.020 - 0.030 in)
Equivalent grades	Champion L – 77J (both models)

1 General description

1 The spark necessary to ignite the petrol/air mixture in the combustion chambers is derived from the alternator attached to the left-hand end of the crankshaft. A twin contact breaker assembly, one set of points for each cylinder, determines the exact moment at which the spark will occur in the cylinder that is due to fire. As the points separate, the low tension circuit is interrupted and a high tension voltage is developed across the points of the spark plug due to fire. This jumps the air gap and ignites the mixture under compression.

2 When the engine is running, the surplus current generated by the alternator is used to provide a regulated 12 volt supply for charging the battery, after it has been converted to direct current by the rectifier. If the battery is fully charged and the demand from the ignition and lighing circuits is low, this excess current is used to reduce the output from the alternator by the appropriate amount.

2 Alternator: checking the output

1 When the ignition switch is turned on, current from the battery flows through the closed points of the voltage regulator to the winding of the alternator rotor, via the positive brush which makes contact with the track on the face of the rotor. Immediately the rotor revolves, the magnetic field of the rotor windings induces current in the three separate coils of the stator assembly surrounding the rotor. This alternating current is used to operate the various electrical circuits and to charge the battery, after rectification to direct current.

2 It follows that good contact must be made between the carbon brushes and the tracks on the end of the alternator rotor with which they make contact. If the tracks or slip rings as they are known become oily or dirty, poor contact will result and the alternator output will drop.

3 The carbon brushes must be free in their holders. The brushes will be found in the stator cover that surrounds the alternator rotor, it also houses the twin contact breaker assembly. When new, the brushes have a length of 11.0 mm (0.433"). The wear limit is 6.0 mm (0.236") when the brushes must be renewed. The brushes and holders are made as an integral unit, simplifying renewal.

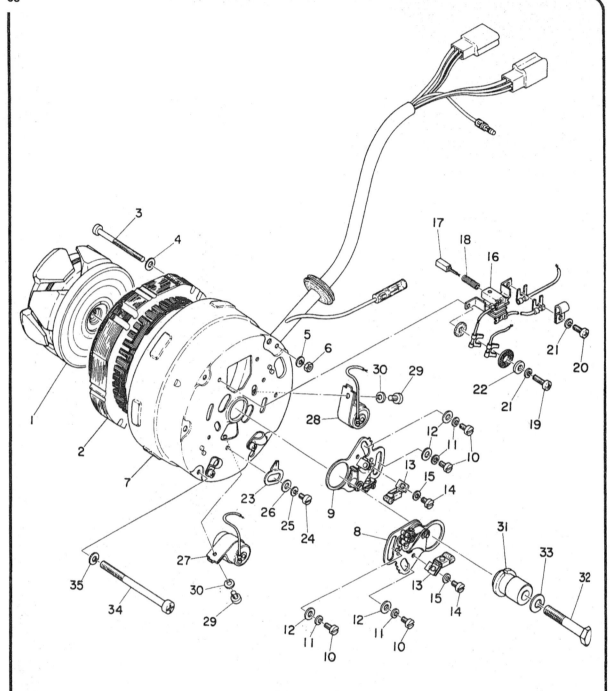

FIG. 3.1. A.C. GENERATOR AND CONTACT BREAKERS

1 Rotor assembly	9 Contact breaker assembly	17 Brush - 2 off	26 Plain washer
2 Stator coil assembly	- right-hand	18 Brush spring - 2 off	27 Condenser - left-hand
3 Screw - 4 off	10 Screw - 4 off	19 Screw	28 Condenser - right-hand
4 Plain washer - 4 off	11 Spring washer - 4 off	20 Screw	29 Screw - 2 off
5 Spring washer - 4 off	12 Plain washer - 4 off	21 Spring washer - 2 off	30 Spring washer - 2 off
6 Nut - 4 off	13 Felt wick lubricator - 2 off	22 Plain washer	31 Cam
7 Stator assembly	14 Screw - 2 off	23 Timing plate	32 Bolt
8 Contact breaker assembly	15 Spring washer - 2 off	24 Screw	33 Spring washer
- left-hand	16 Brush holder assembly	25 Spring washer	34 Stator screw - 3 off
			35 Spring washer - 3 off

Electrode gap check - use a wire type gauge for best results

Electrode gap adjustment - bend the side electrode using the correct tool

Normal condition - A brown, tan or grey firing end indicates that the engine is in good condition and that the plug type is correct

Ash deposits - Light brown deposits encrusted on the electrodes and insulator, leading to misfire and hesitation. Caused by excessive amounts of oil in the combustion chamber or poor quality fuel/oil

Carbon fouling - Dry, black sooty deposits leading to misfire and weak spark. Caused by an over-rich fuel/air mixture, faulty choke operation or blocked air filter

Oil fouling - Wet oily deposits leading to misfire and weak spark. Caused by oil leakage past piston rings or valve guides (4-stroke engine), or excess lubricant (2-stroke engine)

Overheating - A blistered white insulator and glazed electrodes. Caused by ignition system fault, incorrect fuel, or cooling system fault

Worn plug - Worn electrodes will cause poor starting in damp or cold weather and will also waste fuel

2.3a Carbon brushes make contact with alternator slip rings

2.3b To remove, unscrew terminal post and clip around connectors

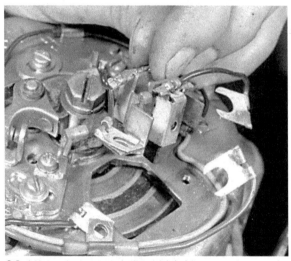

2.3c Lift brushes away complete with holders

3.3a Move fixed contact to adjust points gap

3.3b Points must be fully open when setting or checking gap

4.6 Oil felt wicks occasionally, to lubricate cam

4 The output from the alternator can be checked only with specialised test equipment of the multi-meter type. It is unlikely that the average owner/rider will have access to this type of equipment or instruction in its use. In consequence, if the performance of the alternator is suspect, it should be checked by a Yamaha agent or an auto-electrician.

3 Contact breakers: adjustment

1 To gain access to the contact breaker assembly remove the three cross head screws securing the circular portion of the left-hand crankcase cover. Remove the cover. The twin contact breakers will be found on the stator cover enclosing the rotor of the crankshaft-driven alternator.

2 Rotate the engine until one set of points is fully open, Examine the faces of the contacts. If they are dirty, pitted or burnt, it will be necessary to remove them for further attention, as described in Section 4 of this Chapter. Repeat this operation for the second set of contact points.

3 The correct contact breaker gap, when the points are fully open, is within the range 0.3 - 0.4 mm (0.012'' - 0.016''). Adjustment is effected by slackening the screw holding the fixed contact breaker point in position and moving the point either closer or further away with a screwdriver inserted between the small upright post and the slot in the fixed contact plate. Make sure the points are fully open when this adjustment is made, or a false reading will result. When the gap is correct, tighten the retaining screw and re-check.

4 Repeat the procedure with the other set of contact breaker points.

4 Contact breaker points: removal, renovation and replacement

1 If the contact breaker points are burned, pitted or badly worn, they should be removed for dressing. If it is necessary to remove a substantial amount of material before the faces can be restored, new contacts should be fitted.

2 To remove the moving contact, slacken the screw and nut at the end of the return spring and remove the circlip from the post on which the contact pivots. The moving contact can now be lifted away complete with the return spring and the fibre heel bearing on the contact breaker cam.

3 To remove the fixed contact, remove the screw and nut that has already been slackened, so that the wires can be detached from the end of the mounting plate. Remove the screw that holds the mounting plate in position and lift away the plate complete with fixed contact.

4 When removing the wires from the fixed contact mounting plate, take particular note of the arrangement of the insulating washers. If they are replaced incorrectly, the points will be isolated electrically causing the ignition circuit to fail completely.

5 The points should be dressed with an oilstone or fine emery cloth. Keep them absolutely square during the dressing operation, otherwise they will make angular contact when they are replaced and will burn away rapidly as a result.

6 Replace the contacts by reversing the dismantling procedure, taking care to position the insulating washers in the correct sequence. Lightly grease the pivot post before replacing the moving contact and check that there is no oil or grease on the surface of the points. Place a few drops of oil on the lubricating wick that bears on the contact breaker cam, so that the surface is kept lubricated.

7 Re-adjust the contact breaker gap to the recommended setting, after verifying that the points are in their fully-open position.

8 Repeat the whole procedure for the other set of contact breaker points.

5 Condensers: location, removal and replacement

1 Condensers are included in the contact breaker circuit to prevent arcing across the contact breaker points as they separate. A condenser is connected in parallel with each set of contact points, and if a fault develops in either, or both condensers ignition failure is liable to occur.

2 If the engine is difficult to start, or if misfiring occurs, it is possible that a condenser is at fault. To check whether a condenser has failed, observe the points whilst the engine is running, after removing the circular portion of the left-hand crankcase cover. If excessive sparking occurs across one set of points and they have a blackened or burnt appearance, it may be assumed the condenser in that circuit is no longer serviceable.

3 The condensers are attached to the inside of the stator cover, which must be removed by unscrewing the three crosshead screws around the periphery holding the cover in position Each is retained by a single screw through the strap soldered to the body of the condenser and by the lead wire attached to the screw and nut passing through the end of the moving contact return spring. Remove the screw and nut so that the terminal end is freed, Because it is impracticable to repair a defective condenser, a new one must be fitted.

4 When the replacement is fitted, refit the stator cover and the circular portion of the left-hand crankcase cover. Note that it is extremely unlikely that both condensers will fail in unison; if total ignition failure occurs the source of the trouble should be sought elsewhere.

6 Condensers: testing

1 Without the appropiate test equipment, there is no alternative means of verifying whether a condenser is still serviceable.

2 Bearing in mind the low cost of the condenser, it is far more satisfactory to check whether it is malfunctioning by direct replacement.

7 Ignition coils: checking

1 Each cylinder has its own ignition circuit and if one cylinder misfires, one half of the complete ignition system can be eliminated immediately. The components most likely to fail in the circuit that is defective are the condenser and the ignition coil since contact breaker faults should be obvious on close examination. Replacement of the existing condenser will show whether the condenser is at fault, leaving by the process of elimination the ignition coil.

2 To check whether either ignition coil is at fault, unscrew the spark plug from the cylinder head and lay it across the fins with the insulating plug cap still attached to the coil. Switch on the ignition and check whether a spark occurs at the plug points when the engine is turned over. If no spark is evident, connect the positive lead of a dc voltmeter with a 0.20 volt range to the brown coloured lead to the coil and earth the negative lead. The voltmeter needle should 'kick' as the engine is rotated, as the contact breaker points open and close and interrupt the low

7.4 Twin ignition coils are located under petrol tank

8.1 Ignition switch fits between speedometer and tachometer

tension circuit. If there is no such response, the fault lies in the coil. If the test is satisfactory, disconnect the positive voltmeter lead from the brown lead of the coil and connect it to the orange or grey lead of the coil. The orange lead will provide the same check for the right-hand cylinder and the grey for the left-hand cylinder. A similar 'kick' should be observed if there is electrical continuity through the coil. If a spark is still not evident across the spark plug points as the engine is rotated, either the spark plug is faulty or there is a break in the high tension windings of the coil itself.

3   The ignition coils are sealed units and it is not possible to effect a satisfactory repair in the event of failure. A new coil must be fitted.

4   The ignition coils are mounted as a pair underneath the petrol tank. They bolt direct to a metal plate across the duplex top frame tubes and face in a rearward direction, parallel to the axis of the machine.

8   Ignition switch

1   The ignition switch is a multi-point switch which also controls the lighting circuits. It is bolted from the underside to the top yoke of the forks and is located immediately in front of the handlebars.

2   The switch is unlikely to malfunction during normal service life of the machine and does not require any maintenance. If an ignition failure occurs and it would appear the switch may be responsible, the voltmeter test described in Section 7.2 will confirm whether the switch is at fault. If the voltmeter does not give a reading when it is connected to the brown lead, with the ignition switched on and the contact breaker points closed, the switch is the source of the trouble, assuming the fuse in the electrical circuit has not blown. Replacement of the switch is the only remedy. Reconnection is easy, on account of the junction box connector used.

9   Ignition timing: checking and setting

1   If the ignition timing is correct, the contact breaker points of the cylinder about to fire must be on the verge of separation when the piston is 2.0 mm (0.079") before top dead centre. An approximate indication of the accuracy of the timing is given by the small pointer in one of the apertures of the stator cover of the alternator. This should line up with a scribe mark on

the face of the rotor when the contact breaker points of the cylinder involved are on the point of separation. It must be stressed that this is only an approximate indication of the accuracy of the setting. Optimum performance depends on timing the engine to a high degree of accuracy, to within $\pm$ 0.1 mm of the recommended setting on both cylinders.

2   To set the ignition timing with accuracy, remove the spark plug from the left-hand cylinder and fit a 14 mm dial gauge adaptor. Install the dial gauge and set it so that the dial shows a zero reading when the piston is **exactly** at top dead centre. Rotate the crankshaft backward (clockwise) check that the points for the left-hand cylinder (orange lead) are within the range 0.3 - 0.4 mm (0.012" - 0.016") when they are fully open, then reverse the direction of rotation until the piston is 2.0 mm **exactly** from top dead centre. If the timing is correct, the contact breaker points should be about to separate.

3   If the timing is not correct, adjust the contact breaker points by moving them as a unit either clockwise or anti-clockwise, depending on whether the opening point needs to be advanced or retarded. This is accomplished by slackening the two cross head screws that pass through the elongated slots in the contact breaker base plate and moving the base plate with a screwdriver blade inserted between the short upright post and the notches in the edge of the base plate. When the adjustment is correct, tighten both screws and re-check the setting.

4   Repeat this procedure for the right-hand cylinder, adjusting the points to which the grey lead is attached. This setting must be made with equal accuracy.

5   As a final check, attach the positive lead of a 0-20 volt range dc voltmeter to each moving contact in turn and earth the negative lead. Switch on the ignition and check that the voltmeter commences to show a reading when the piston is 2.0 mm from top dead centre. Repeat for the other cylinder, with the other set of contact breaker points.

10   Spark plugs: checking and re-setting the gap

1   Two NGK 14 mm spark plugs are fitted to the Yamaha 250/350 cc twins as standard equipment. The B-8HCS grade is recommended for the 250 cc YDS7 model and the B-9HC grade for the 350 cc YR5 model. Both grades of plug have a ½ inch reach.

2   The recommended spark plug gap is 0.5 — 0.7 mm (0.020 — 0.030 in). Check the gap every 1,000 miles. To re-set, bend the outer electrode away from or closer to the centre electrode and check that a 0.025" feeler gauge can be inserted. Never bend the central electrode, otherwise the insulator will crack, causing engine damage if the broken particles fall in whilst the engine is running.

3   After some experience the spark plug electrodes can be used as a reliable guide to engine operating conditions. See accompanying diagrams.

4   Always carry two spare spark plugs of the correct type. The plugs in a two-stroke engine lead a particularly hard life and are liable to fail more readily than when fitted to a four-stroke.

5   Never overtighten a spark plug, otherwise there is risk of stripping the threads from the cylinder head, especially as it is cast in light alloy. A stripped thread can be repaired without having to scrap the cylinder head by using a 'Helicoil' thread insert. This is a low-cost service, operated by a number of dealers.

6   Use the correct size spanner when tightening plugs, otherwise the spanner may slip and damage the ceramic insulators. The plugs should be tightened sufficiently to seat firmly on their sealing washers, and no more.

7   Make sure that the plug insulating caps are a good fit and free from cracks. Apart from acting as an insulator from water and road dirt they contain the suppressor for eliminating radio and TV interference.

9.3a Slacken screws in base plate ...

9.3b ... Use screwdriver to reposition points, before tightening

11 Fault diagnosis - Ignition system

Symptom	Reason/s	Remedy
Engine will not start	No spark at plugs	Faulty ignition switch. Check whether current is reaching ignition coils.
	Weak spark at plugs	Dirty contact breaker points require cleaning. Contact breaker gaps have closed up. Re-set.
Engine starts, but runs erratically	Intermittent or weak spark on one cylinder	Locate defective cylinder and replace plug. If no improvement check whether points are arching. If so replace condenser.
	Ignition over-advanced	Check ignition timing and if necessary, re-set
	Plug lead insulation breaking down	Check for breaks in outer covering, especially near frame.
Engine difficult to start and runs sluggishly. Overheats.	Ignition timing retarded	Check ignition timing and advance to correct setting.

Chapter 4 Frame and forks

Contents

1 General description

The Yamaha 250 cc and 350 cc twins employ a common frame and fork assembly of conventional design. The front forks are telescopic, with oil-filled one-way damper units. The frame is of the full cradle type, employing duplex tubes. Rear suspension is provided by a swinging arm fork with replaceable bushes, controlled by hydraulically clamped, adjustable, rear suspension units.

2 Front forks: removal from the frame

1 It is unlikely that the front forks will need to be removed from the frame as a complete unit unless the steering head bearings require attention or the forks are damaged in an accident.
2 Start by removing either the control cables from the handlebar control levers or the levers complete with cables. The shape of the handlebars and the length of the control cables will probably dictate which method is used.
3 Detach the handlebars together with their mountings. The mountings pass through rubber bushes in the top yoke of the forks, which act as a vibration damper. They are retained by a nut, plain washer and spring washer on the underside, which can be removed after the split pin passing through a drilling in each handlebar mounting has been withdrawn.
4 It will be necessary to remove the steering damper knob before the handlebars and their mountings can be freed from the fork yoke. Remove the split pin from the extreme end of the rod attached to the steering damper rod and unscrew the damper knob until it lifts clear with the rod attached. Detach the headlamp complete, which is held to the upper fork shrouds by two bolts that thread into the headlamp shell.
5 If the machine is not already resting on the centre stand, support it in this manner on firm, level ground. Balance the machine so that the front wheel is clear of the ground and place some packing under the crankcase so that if the machine should inadvertently tip forward, it will not roll off the centre stand.
6 Remove the speedometer drive cable from the drive within the front hub by releasing the spring clip and withdrawing the cable. Detach the front brake cable by removing the split pin through the clevis pin passing through the brake operating arm. When the clevis pin is withdrawn, the cable can be detached complete with rubber gaiter and cable adjuster, after the latter has been unscrewed from the brake plate.
7 The front wheel can now be released by withdrawing the spindle, which passes through the left-hand fork leg and is retained by a castellated nut, and split pin. Note that it will be necessary to slacken the two nuts which secure the clamp around the head of the spindle, at the extreme end of the right-hand fork leg. The head of the spindle is drilled, so that a tommy bar can be inserted, to aid removal. The wheel will pull clear after the anchorage slot on the brake plate has disengaged from the abutment on the left-hand fork leg.
8 If desired, the front mudguard can be removed at this stage. It is secured to the inside of each fork leg by two bolts and washers which, when removed, will release the mudguard and stays as a complete unit. The speedometer cable and the front brake cable will pass through the guide that keeps them clear from the front wheel.
9 Unscrew the large nut in the centre of the top yoke of the forks and remove it together with the steering damper spring and the crown washer. Detach the drive cable from both the tachometer and the speedometer heads. Remove both instrument heads together with their rubber mountings by unscrewing the nut securing each mounting pla te to the fork yoke.

10 Slacken the pinch bolt at the rear of the top yoke and remove both chromium plated bolts from the top of each fork leg. The yoke can now be lifted away, if necessary by lightly tapping the underside with a rawhide mallet to free it initially.
11 Whilst the forks are supported in this position it is a convenient opportunity to drain off the damping oil, especially if further dismantling is necessary after the forks have been removed from the frame. The drain screws (cross head) are found at the lower end of each fork leg; place a container below each to catch the oil when the screw is removed and discard the oil.
12 Unscrew the slotted nut at the head of the steering column, using a 'C' spanner of the correct shape. As the nut is slackened, the forks will gradually ease away from the steering head, uncovering the uncaged ball bearings of the steering head races. Make provision to catch the ball bearings as they are released; only the lower bearings will drop free since the upper bearings will most probably remain seated in the cup retaining them.
13 When the slotted nut has been removed from the steering

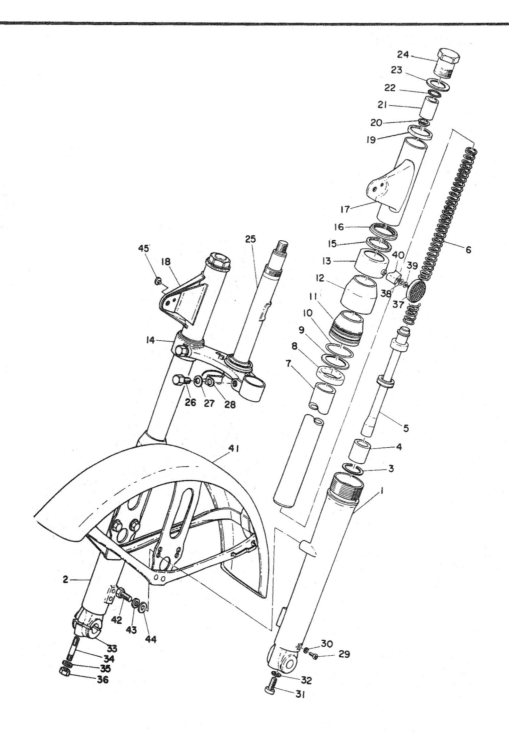

FIG. 4.1. FRONT FORKS AND MUDGUARD – COMPONENT PARTS

1 Left-hand lower fork leg	12 Dust seal cover - 2 off	23 Cap washer - 2 off	34 Stud - 2 off
2 Right-hand lower fork leg	13 Outer cover - left-hand	24 Cap bolt - 2 off	35 Spring washer - 2 off
3 Circlip - 2 off	14 Outer cover - right-hand	25 Lower fork yoke complete	36 Nut - 2 off
4 Damping piston - 2 off	15 Packing - 2 off	26 Pinch bolt - 2 off	37 Reflector - 2 off
5 Damper unit complete - 2 off	16 Cover guide - 2 off	27 Spring washer - 2 off	38 Plain washer - 2 off
6 Fork spring - 2 off	17 Upper fork shroud - left-hand	28 Cable holder	39 Spring washer - 2 off
7 Fork stanchion - 2 off	18 Upper fork shroud - right-hand	29 Drain plug - 2 off	40 Reflector mounting - 2 off
8 Oil seal - 2 off	19 Upper fork shroud guide -	30 Drain plug washer	41 Front mudguard complete
9 Oil seal washer - 2 off	2 off	31 Bolt - 2 off	42 Bolt - 4 off
10 Oil seal clip - 2 off	20 Upper spring seat - 2 off	32 Packing - 2 off	43 Spring washer - 4 off
11 Dust seal - 2 off	21 Spacer - 2 off	33 Split clamp	44 Plain washer - 4 off
	22 Packing - 2 off		45 Blanking off plug - 2 off

column completely, the fork can be withdrawn from the lower end of the steering head as a complete unit. It may be necessary to raise the machine even higher during this operation, so that the fork stem will clear the steering head.

3 Front forks: dismantling

1 The fork legs can be dismantled individually, without need to disturb the steering head bearings. The preliminary dismantling is accomplished by following the procedure detailed in paragraphs 5-8 of the preceding Section, then continuing with the instructions given in this Section, after removing the chromium plated bolts from the top of each fork leg and draining off the damping oil.

2 If both fork legs are to be dismantled, strip them separately, using an identical procedure. There is less chance of unwittingly interchanging parts if this approach is adopted.

3 Slacken the pinch bolt through the lower fork yoke and pull the complete fork leg from the assembly, leaving the upper fork shroud in position. Invert the fork with the spring still in position and using an hexagonal allen key, remove the socket screw recessed into the curved portion of the lower fork end, through which the wheel spindle normally passes. The spring pressure is essential, to prevent the fork damper unit from rotating whilst this socket screw is removed.

4 Remove the top bush and spring from within the fork leg, then re-invert the fork leg and prise off the chromium plated dust cover fitting over the dust seal, where the sliding action of the fork occurs. Remove the dust seal and the circlip within the top end of the lower fork leg retaining the oil seal and oil seal washer. The fork tube can now be pulled away from the lower fork leg and separated.

5 To remove the damper unit, invert the fork leg and detach the circlip inside the tapered bottom end. The damper assembly, complete with piston, can now be drawn out as a complete assembled unit. No further dismantling is possible.

4 Steering head bearings: examination and renovation

1 Before reassembly of the forks is commenced, examine the steering head races. The ball bearing tracks of the respective cup and cone bearings should be polished and free from indentations or cracks. If wear or damage is evident, the cups and cones must be renewed as a complete set. They are a tight press fit and should be drifted out of position.

2 Ball bearings are cheap. If the originals are marked or discoloured, they should be renewed. To hold the steel balls in position during reassembly, pack the bearings with grease. Note that each race contains only nineteen ¼" ball bearings. There is space for the addition of one extra ball, but this must be left empty to prevent the ball bearings from skidding on one another, a situation which would greatly accelerate the rate of wear.

5 Front forks: examination and renovation

1 The parts most liable to wear over an extended period of service are the bush which fits immediately below the chromium plated bolt at the top of the fork leg, the damper assembly within the fork tube and the oil seal at the sliding joint. Wear is normally accompanied by a tendency for the forks to judder when the front brake is applied and it should be possible to detect the increased amount of play by pulling and pushing on the handlebars when the front brake is applied fully. This type of wear should not be confused with slack steering head bearings, which can give identical results.

2 Renewal of the worn parts is quite straightforward. Particular care is necessary when renewing the oil seal, to ensure that the feather edge seal is not damaged during reassembly. Both the seal and the fork tube should be greased, to lessen the risk of

2.6a Detach speedometer cable by removing retaining clip

2.6b Disconnect front brake cable by withdrawing clevis pin

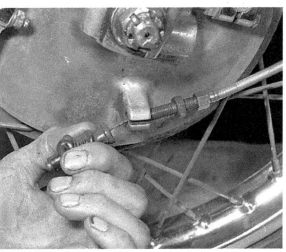

2.6c then unscrew adjuster and remove from brake plate

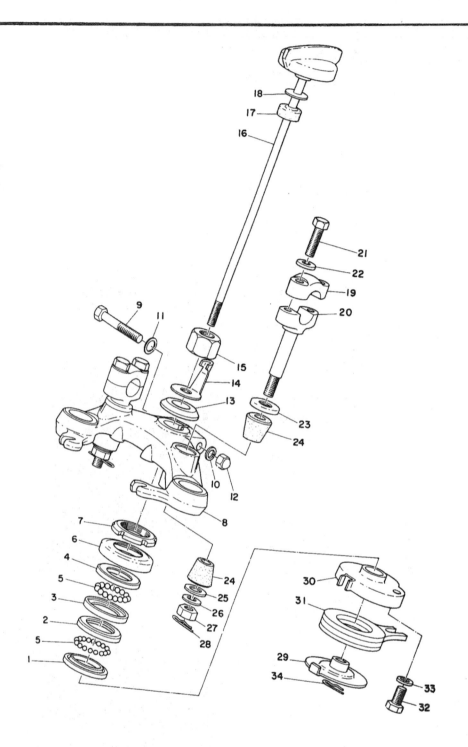

FIG. 4.2. STEERING DAMPER — COMPONENT PARTS

1 *Lower steering head cone*	9 *Bolt*	18 *Washer*	27 *Nut - 2 off*
2 *Lower steering head cup*	10 *Spring washer*	19 *Split clamp - 2 off*	28 *Clip - 2 off*
3 *Upper steering head cup*	11 *Washer*	20 *Lower clamp - 2 off*	29 *Damper plate - lower*
4 *Upper steering head cone*	12 *Acorn nut*	21 *Bolt - 4 off*	30 *Damper plate - upper*
5 *Ball bearings, ¼ in. diam.*	13 *Crown washer*	22 *Spring washer - 4 off*	31 *Damper plate*
- 38 off	14 *Damper spring*	23 *Plain washer - 2 off*	32 *Bolt*
6 *Ball race cover*	15 *Nut*	24 *Rubber bush - 4 off*	33 *Spring washer*
7 *Adjusting nut*	16 *Damper rod*	25 *Washer - 2 off*	34 *Clip*
8 *Upper fork yoke*	17 *Washer*	26 *Spring washer - 2 off*	

2.7 Wheel spindle pulls out from right

2.8 Front mudguard is bolted to inside of fork legs

2.10 Remove plated bolts from top of each fork tube

2.11 Remove drain plugs and drain off oil

damage.

3   After an extended period of service, the fork springs may take a permanent set. The free length is 389.5 mm and it is wise to fit new components if the overall length has decreased. Always fit new springs as a pair, NEVER separately.

4   Check the outer surface of the fork tube for scratches or roughness. It is only too easy to damage the oil seal during reassembly, if these high spots are not eased down. The fork tubes are unlikely to bend unless the machine is damaged in an accident. Any significant bend will be detected by eye, but if there is any doubt about straightness, roll the tubes on a flat surface. If the tubes are bent, they must be renewed. Unless specialised repair equipment is available, it is rarely practicable to straighten them to the necessary standard.

5   The dust seals must be in good order if they are to fulfill their proper function. Replace any that are split or damaged.

6   Damping is effected by the damper units contained within each fork tube. The damping action can be controlled within certain limits by changing the viscosity of the oil used as the damping medium, although a change is unlikely to prove necessary except in extremes of climate.

6   Front forks: replacement

1   Replace the front forks by reversing either of the dismantling procedures described in Sections 2 and 3 of this Chapter, whichever the more appropriate. Make sure the abutment of the brake plate aligns correctly with the retaining slot cast in the left-hand lower fork leg.

2   Before fully tightening the front wheel spindle, right-hand spindle clamp, fork yoke pinch bolts and the chromium plated bolts in the top of each fork leg, bounce the forks several times to ensure that they work freely and settle down in their original positions. Complete the final tightening from the front wheel spindle upward, and do not forget to fit and open the split pin through the wheel spindle nut.

3   Do not forget to add the correct amount of damping oil to each fork leg before the bolts in the top are tightened. Each fork should be filled with 145 cc SAE 10W/30 engine oil. Check that the drain plugs at the front of each fork leg have been replaced and tightened, before the oil is added!

4   Difficulty is often experienced when attempting to draw the fork inner tube into position in the top yoke during assembly, even though the tube does not have a taper fit. A Yamaha service tool is available for this purpose, in the form of a portion of threaded rod on which a 'T' handle has been brazed across the top. The rod screws into the thread on the inside of the fork tube and can be used as a guide to draw the tube into position. It is easy

3.3a Slacken pinch bolt in lower yoke to free fork leg

3.3b Complete fork leg can be pulled from fork assembly

3.3c Invert, with spring in position, then ...

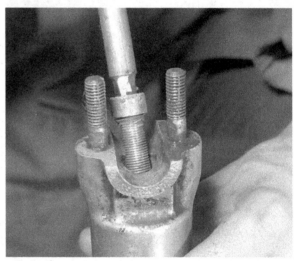

3.3d Unscrew fork damper socket screw from fork leg

3.4a Remove top fork bush and also ...

3.4b ... fork spring, which is a loose fit

3.4c Prise off dust cover, then pull off ...

3.4d ... then dust seal, which will expose ...

3.4e ... circlip within lower fork leg

3.4f Fork tube can now be pulled out of lower leg

3.5a Remove circlip in base of tube to free damper assembly

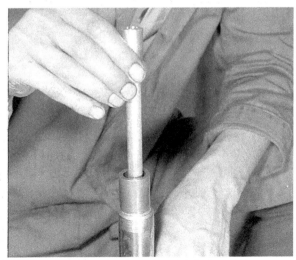

3.5b Damper unit can now be drawn out of tube

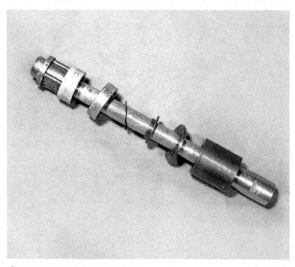

3.5c The complete fork damper assembly

6.1 Brake plate must line up so that anchorage engages correctly

6.2 Do not forget the split pin through the front spindle nut

to construct a similar tool or if time is short, to thread the tapered end of the broom handle into the fork tube as a temporary expedient.

5 Check the adjustment of the steering head bearings before the machine is used on the road and again shortly afterwards, when they settle down. If the bearings are too slack, fork judder will occur. There should be no play at the headraces when the handlebars are pulled and pushed hard, with the front brake applied hard.

6 Overtight headraces are equally undesirable. It is possible to place a pressure of several tons on the head bearings by over-tightening, even though the handlebars may seem to turn quite freely. Overtight bearings will cause the machine to roll at low speeds and give imprecise steering. Adjustment is correct if there is no play in the bearings and the handlebars swing to full lock either side when the machine is on the centre stand with the front wheel clear of the ground. Only a light tap on each end should cause the handlebars to swing.

7 Steering head lock

1 The steering head lock is attached to the left-hand side of the steering head. It is retained by a rivet. When in a locked position, the plunger extends and engages with a portion of the steering head stem, so that the handlebars are locked in position and cannot be turned.

2 If the lock malfunctions, it must be renewed. A repair is impracticable. When the lock is changed it follows that the key must be changed too, to correspond with the new lock.

8 Steering damper: function and use

1 A steering damper is a fitting providing a means of adding friction to the steering head assembly so that the front forks will turn less easily. It is a relic of the early days of motor cycling when machines were prone to develop 'speed wobbles' which grew in intensity and eventually unseated the rider. With today's more sophisticated front form damping and improved frame design, a steering damper is virtually a superfluous fitting unless a sidecar is attached or very poor road surfaces are encountered.

2 The steering damper can be likened to a small clutch without compression springs. When the steering damper knob is tightened, the friction discs and the plain discs of the assembly are brought into closer proximity and it is more difficult to deflect the handlebars from their set position. When the knob is tightened fully, the handlebars are locked in position. Under normal road conditions, the steering damper should be slackened off. Only at very high speeds or on rough surfaces is there any need to apply some damper friction.

3 The steering damper assembly will be found at the base of the steering head column, immediately below the bottom fork yoke. The centre fixed plate is attached to the fork yoke by a bolt and washer.

4 Although it is unlikely that the steering damper assembly will require attention during the normal service life of the machine, it can be removed by releasing the split pin through the extreme end of the rod attached to the steering damper knob, as described in Section 2.4 of this Chapter. The remainder of the assembly is freed when the bolt retaining the centre plate in position is withdrawn from the underside of the bottom fork yoke.

9 Frame: examination and renovation

1 The frame is unlikely to require attention unless it is damaged as the result of an accident. In many cases, replacement of the frame is the only satisfactory course of action, if it is badly out of alignment. Comparatively few frame repair special-ists have the necessary mandrels and jigs essential for the accurate re-setting of the frame and, even then there is no means

of assessing to what extent the frame may have been overstressed such that a later fatigue failure may occur.

2   After a machine has covered an extensive mileage, it is advisable to keep a close watch for signs of cracking or splitting at any of the welded joints. Rust can cause weakness at these joints particularly if they are unpainted. Minor repairs can be effected by welding or brazing, depending on the extent of the damage found.

3   A frame out of alignment, will cause handling problems and may even promote 'speed wobbles' in a particular speed range. If misalignment is suspected as the result of an accident, it will be necessary to strip the machine so that the frame can be checked, and if needs be, renewed.

10 Swinging arm rear fork: dismantling, examination and renovation

1   The rear fork of the frame assembly pivots on a detachable bush within each end of the fork cross member and a pivot shaft which passes through frame lugs and the centre of each of the two bushes. It is quite easy to renovate the swinging arm pivots when wear necessitates attention.

2   To remove the swinging arm fork, first place the machine on the centre stand, then detach the final drive chain, preferably whilst the spring link is resting in the teeth of the rear wheel sprocket. It is advisable to remove the chainguard, which is attached at the rear by a cross head screw that threads into the left-hand chainstay, immediately to the rear of the lower end of the suspension unit mounting. The forward end of the chainguard is attached to the left-hand side of the swinging arm fork by a cross head screw which passes through a rubber mounting. The screw head is in close proximity to the rectifier which is easily damaged it may be advisable to leave the chainguard in position if the screw will not remove easily. The swinging arm fork can be detached quite satisfactorily with the chainguard still in position.

3   Detach the rear brake torque arm at the brake plate by removing the split pin, then unscrewing and detaching the nut and washer. Unscrew and remove the brake rod adjusting nut and free the rod from the brake operating arm, taking care not to lose the trunnion through which the rod passes and the spring around the rod. Remove the split pin from the left-hand end of the rear wheel spindle, then the castellated nut. The rear wheel spindle can now be unscrewed and withdrawn completely leaving the wheel free to be removed from the frame after the large nut has been removed from the sprocket shaft. Do not lose the spacer on the right-hand side of the hub, through which the wheel spindle passes. The chain tensioners will most probably pull away with the wheel and fall free.

4   Remove both rear suspension units. Each is required in position by a domed nut and washer. When the nuts and washers have been removed, the suspension units can be pulled off their mounting studs.

5   Separate the locknuts on the right-hand side of the pivot shaft which passes through the frame gussets and withdraw them from the shaft. Detach the stop light switch. Pull the pivot shaft away from the left-hand side of the machine, then pull the swinging arm fork from the rear. It will come away complete with the thrust covers and any shims that have been added to take up play. Note the position of the shims and their thickness so that they are replaced in the correct order.

6   Remove both thrust covers and withdraw the bushes from each end of the cross member. Note that there is a much longer distance piece between them which serves no load bearing function.

7   Wear will take place in both the bushes and the pivot shaft, which should be renewed together, never separately. It is also advisable to renew the shaft if it is out of true, irrespective of the condition of the bushes.

8   Reassemble the swinging arm fork by reversing the dismantling procedure. Grease the pivot shaft and the bearings liberally prior to assembly and check that the grease nipples in both ends of the pivot shaft are unobstructed. When reassembly

6.3 Damping oil must be added before fork top bolts are replaced

6.5 Turn slotted nut to adjust steering head bearings

7.1 Steering head lock is rivetted in position

10.2 Detach chain at rear wheel sprocket

10.3a Detach rear brake torque arm at brake plate

10.3b Unscrew adjuster from brake rod

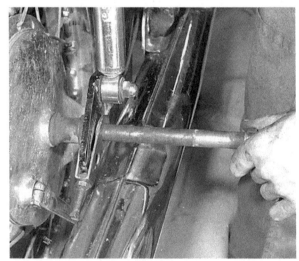
10.3c Pull out wheel spindle after removing nut

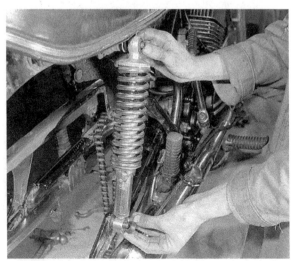
10.4 Remove both rear suspension units

10.5a Unscrew the two locknuts on the pivot shaft

10.5b Remove also the stop lamp switch

10.5c Pull away the swinging arm after the pivot shaft is withdrawn

10.5d Shims may be used to take up side play

10.6a Bushes will pull out of each end

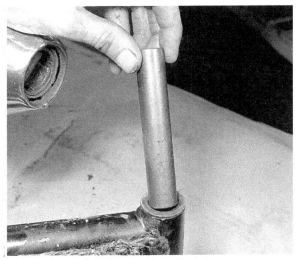

10.6b Long distance piece separates bushes

10.8a Grease pivot shaft prior to insertion

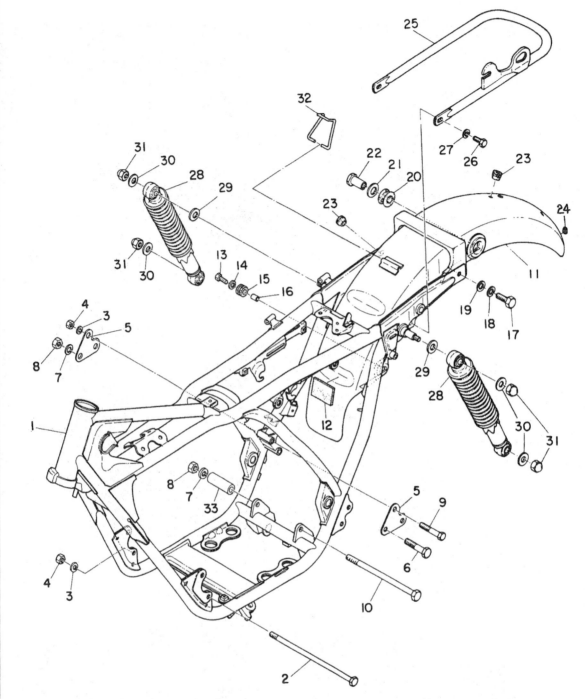

FIG. 4.3. FRAME, REAR SUSPENSION UNITS AND REAR MUDGUARD – COMPONENT PARTS

1 Frame unit complete	12 Battery box seal	23 Tail lamp grommet - 2 off
2 Engine bolts - front - 2 off	13 Bolt - 2 off	24 Damper stop
3 Spring washer - 4 off	14 Washer - 2 off	25 Rear mudguard stay
4 Nut - 4 off	15 Battery box damper - 2 off	26 Bolt - 2 off
5 Engine rear upper stay - 2 off	16 Mudguard mounting collar - 2 off	27 Spring washer - 2 off
6 Bolt	17 Bolt - 2 off	28 Rear suspension units - 2 off
7 Spring washer - 2 off	18 Spring washer - 2 off	29 Washer for rear suspension units (lower) - 2 off
8 Nut - 2 off	19 Plain washer - 2 off	
9 Bolt - 2 off	20 Rear damper - 2 off	30 Washer for rear suspension units (upper) - 2 off
10 Bolt	21 Washer - 2 off	
11 Rear mudguard complete	22 Rear mudguard nut - 2 off	31 Acorn nut - 4 off
		32 Wiring harness clamp
		33 Engine mounting distance piece

is complete, apply a grease gun to both ends of the pivot shaft and continue pumping until grease can be seen emerging from the ends of the pivot joint.

9 Apart from causing a machine MOT failure, worn swinging arm pivot bearings will give imprecise handling, with a tendency for the rear of the machine to twitch or hop. The play can be detected by placing the machine on the centre stand with the rear wheel clear of the ground, by pulling and pushing alternately on the fork ends.

11 Rear suspension units: examination

1 Rear suspension units of the hydraulically-damped type are fitted to the 250 cc and 350 cc Yamaha twins. They can be adjusted to give three different spring loadings, without removal from the machine.

2 Each rear suspension unit has two peg holes immediately above the adjusting notches, to facilitate adjustment. Either a 'C' spanner or the screwdriver supplied with the original tool kit can be used to turn the adjusters. Turn clockwise to increase the spring tension and stiffen up the rear suspension.

The recommended settings are:-

Position 1 (least tension)   Normal solo riding
Position 2 (middle setting) High speed touring
Position 3 (greatest tension) High speed competition events or with pillion passenger and/or heavy loads.

3 The suspension units are sealed and there is no means of topping up or changing the damping fluid. If the damping fails or if the unit leaks, renewal is necessary.

4 In the interests of good roadholding it is essential that both suspension units have the same load setting. If a renewal is necessary, the units must be replaced as a matched pair.

12 Centre stand: examination

1 The centre stand is attached to lugs welded to the bottom of the rearmost cross member of the duplex tube frame, below the engine. The pivot is bushed and the stand is retained by a pivot shaft with a split pin through one end. An extension spring, fitted on the right-hand side, keeps the stand in the fully-retracted position when the machine is in use.

2 Check that the return spring is in good condition and correctly located. If the stand drops when the machine is in motion it may catch in some obstacle and unseat the rider.

13 Prop stand: examination

1 A prop stand is fitted, for occasional parking when it is not desired to use the centre stand. The prop stand pivots from a lug welded to the lower left-hand tube, close to the crankcase of the engine. A bolt and nut pass through both the prop stand and the lug to act as the pivot; the nut is retained by a split pin that passes through a drilling in the end of the bolt. An extension spring returns the stand to the retracted position, immediately the weight is taken from the prop stand.

2 Check that the split pin has not been omitted from the pivot bolt and that the nut is tight. Check also that the extension spring is not overstretched or worn at the end connection. An accident is almost inevitable if the prop stand should fall whilst the machine is on the move.

14 Footrests: examination and renovation

1 The footrests are made as a complete unit, which is held to lugs welded to the underside of the duplex frame tubes below the engine by four nuts and bolts. The nuts are retained by split pins that pass through drillings in the ends of the bolts; rubber

10.8b Pack whole assembly with grease after insertion

anti-vibration mountings give the footrests a certain amount of flexibility.

2 If the machine is dropped, it is probable that one or other of the footrest arms will bend as the result of impact with the road. If the complete footrest assembly is removed from the machine and the footrest rubbers taken off, the damaged arm can be straightened in a vice, using a blowlamp flame to apply heat at the area where the bend occurs.

15 Rear brake pedal: examination and renovation

1 The rear brake pedal pivots through the right-hand silencer stay, to which a short length of tube is welded. The shaft carrying the brake arm is splined, to engage with splines of the rear brake pedal. The pedal is retained to the shaft by a simple pinch bolt arrangement.

2 If the brake pedal is bent or twisted in an accident, it should be removed by slackening the pinch bolt and straightened in a manner similar to that recommended for the footrests in the preceding Section.

3 Make sure the pinch bolt is tight. If the lever is a slack fit on the splines, they will wear rapidly and it will be difficult to keep the lever in position.

16 Kickstarter lever: examination and renovation

1 The kickstarter lever is splined and is secured to its shaft by means of a pinch bolt. The kickstarter crank swivels so that it can be tucked out of the way when the engine is started. It is held in position on the swivel by a washer and circlip. A spring-loaded ball bearing locates the kickstarter arm in either the operating or folded position; if the action becomes sloppy it is probable that the spring behind the ball bearing needs renewing. It is advisable to remove the circlip and washer occasionally, so that the kickstarter crank can be detached and the swivel greased.

2 It is unlikely that the kickstarter crank will bend in an accident unless the machine is ridden with the kickstarter in the operating and not folded position. It should be removed and straightened, using the same technique as that recommended for the footrests in Section 14.2.

17 Dual seat: removal and replacement

1 The dual seat is attached to the right-hand frame tube by means of a pivot on which it hinges. A catch on the left-hand

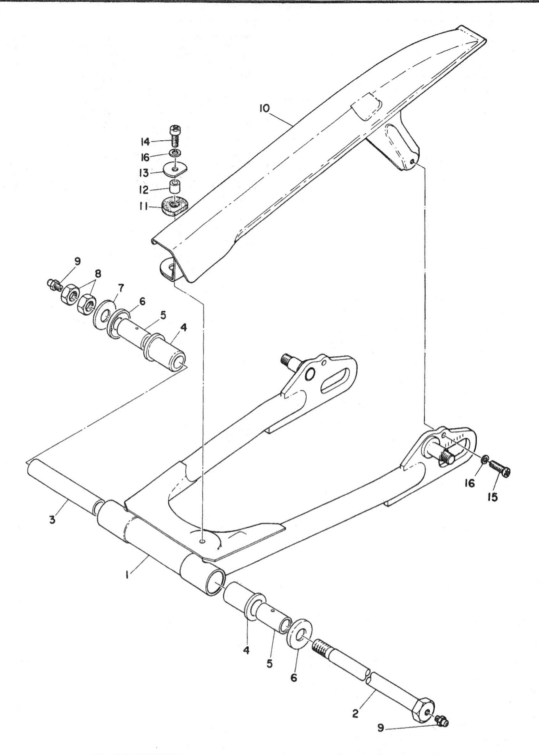

FIG. 4.4. SWINGING ARM FORK AND CHAINGUARD — COMPONENT PARTS

1 Swinging arm fork complete
2 Pivot shaft
3 Distance collar
4 Outer bushes - 2 off
5 Inner bushes - 2 off
6 Thrust cover - 2 off
7 Shim - number as required
8 Pivot shaft nut

9 Grease nipple
10 Chainguard
11 Chainguard damper
12 Mudguard mounting collar
13 Chainguard washer
14 Screw
15 Screw
16 Spring washer - 2 off

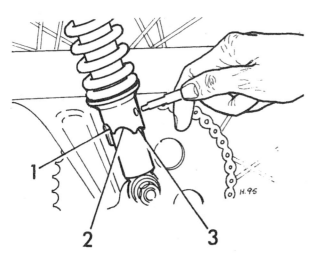

Fig. 4.5. Settings for rear suspension units

frame tube locks the dual seat in position, under normal riding conditions.

2 To release the dual seat from the machine, lift the catch and lift the dual seat so that the pivot on the right-hand side is exposed. If the split pin though the pivot pin is removed and the pivot pin withdrawn, the dual seat can be lifted away.

18 Speedometer and tachometer heads: removal and replacement

1 The speedometer and tachometer heads are rubber-mounted and attached to the top fork yoke by means of a single bolt fixing. If the bolt is slackened and the drive cable detached, the head complete with mounting can be lifted away; the yoke is slotted to aid removal. It will be necessary to remove the bulbs from the case of each instrument head by pulling the bulb-holders from their seatings; each is retained by a rubber cup.

2 The rubber mountings are retained to each instrument case by two split pins which pass through two short columns attached to the base of the instrument case. Do not misplace the rubber cushion interposed between the mounting bracket and the instrument case to damp out the undesirable effects of vibration.

3 Apart from defects in either the drive or the drive cable, a speedometer or tachometer that malfunctions is difficult to repair. Fit a new one, or alternatively entrust the repair to a component instrument repair specialist.

4 Remember that a speedometer in correct working order is a stutatory requirement in the UK. Apart from this legal requirement, reference to the odometer reading is the best means of keeping in pace with the maintenance schedule.

19 Speedometer and tachometer drive cables: examination and maintenance

1 It is advisable to detach both cables from time to time in order to check whether they are lubricated adequately, and whether the outer coverings are compressed or damaged at any point along their run. Jerky or sluggish movements can often be attributed to a cable fault.

2 For greasing, withdraw the inner cable. After wiping off the old grease, clean with a petrol-soaked rag and examine the cable for broken strands or other damage.

3 Regrease the cable with high melting point grease, taking care not to grease the last six inches at the point where the cable enters the instrument head. If this precaution is not observed, grease will work into the head and immobilise the movement.

4 If either instrument ceases to function, suspect a broken cable. Inspection will show whether the inner cable has broken; if so, the inner cable alone can be renewed and reinserted in the outer casing, after greasing. Never fit a new inner cable alone if the outer covering is damaged or compressed at any point.

20 Speedometer and tachometer drive: location and examination

1 The speedometer drive gearbox is an integral part of the front wheel brake plate and is driven internally from the wheel hub. The gearbox rarely gives trouble if it is lubricated whenever the front wheel and brake plate are removed; there is no external grease nipple. If wear in the drive mechanism occurs, the worm complete with shaft can be withdrawn from the brake plate housing by unscrewing a pegged bush. The drive pinion is retained to the inside of the brake plate by a circlip, in front of the shaped driving plate that takes up the drive from the wheel hub.

2 The tachometer drive is taken from the gear pinion, integral with the clutch outer drum, via the kickstarter idler pinion. It is unlkely that the drive will give trouble during normal service life of the machine.

21 Cleaning the machine

1 After removing all surface dirt with a rag or sponge washed frequently in clean water, the machine should be allowed to dry thoroughly. Application of car polish or wax to the cycle parts will give a good finish, particularly if the machine has been neglected for a long period.

2 The plated parts of the machine should require only a wipe with a damp rag. If the plated parts are badly corroded, as may occur during the winter when the roads are salted, it is preferable to use one of the proprietary chrome cleaners. These often have an oily base, which will help to prevent the corrosion from re-occurring.

3 If the engine parts are particularly oily, use a cleaning compound such as "Gunk" or "Jizer". Apply the compound whilst the parts are dry and work it in with a brush so that it has the opportunity to penetrate the film of grease and oil. Finish off by washing down liberally with plenty of water, taking care that it does not enter the carburettors or the electrics. If desired, the now clean aluminium alloy parts can be enhanced further by using a special polish such as Solvol "Autosol", which will fully restore their brilliance.

4 Whenever possible, the machine should be wiped down after it has been used in the wet, so that it is not garaged under damp conditions which will promote rusting. Make sure to wipe the chain and re-oil it, to prevent water from entering the rollers and causing harshness with an accompanying high rate of wear. Remember there is little chance of water entering the control cables and causing stiffness of operation if they are lubricated regularly as recommended in the Routine Maintenance Section.

22 Fault diagnosis — Frame and forks

Sympton	Reason/s	Remedy
Machine veers either to the left or the right with hands off handlebars	Bent frame Twisted forks Wheels out of alignment	Check, and replace. Check, and replace. Check and realign.
Machine rolls at low speed	Overtight steering head bearings	Slacken until adjustment is correct.
Machine judders when front brake is applied	Slack steering head bearings Worn fork bushes	Tighten, until adjustment is correct. Dismantle forks and replace bushes.
Machine pitches on uneven surfaces	Ineffective fork dampers Ineffective rear suspension units Suspension too soft	Check oil content. Check whether units still have damping action. Raise suspension unit adjustment one notch.
Fork action stiff	Fork legs out of alignment (twisted in yokes)	Slacken yoke clamps, and fork top bolts. Pump fork several times then retighten from bottom upwards.
Machine wanders. Steering imprecise. Rear wheel tends to hop.	Worn swinging arm pivot	Dismantle and replace bushes and pivot shaft.

Chapter 5 Wheels, brakes and tyres.

Contents

Specifications

Tyres

									YDS7	YR5
Front ...	...	...	...	...	...	...	...	...	3.00 x 18 in	3.00 x 18 in
Rear ...	...	...	...	...	...	...	...	...	3.25 x 18 in	3.50 x 18 in

Tyre pressures:										
Front	...	...	...	...	...	...	...	...	23 psi	23 psi
Rear	...	...	...	...	...	...	...	...	29 psi	29 psi

Note: For continuous high speed riding, or when a pillion passenger is carried, add 5 psi to the above recommendations, for both tyres.

Brakes

Front ...	...	...	...	...	...	...	...	...	Twin leading shoe, 8 inch diameter
Rear ...	...	...	...	...	...	...	...	...	Single leading shoe, 8 inch diameter

1 General description

1 Both wheels are of 18 inch diameter. They carry a ribbed tread tyre of 3.00 inch section on the front wheel and in the case of the 250 cc YDS7 model, a block tread rear tyre of 3.25 inch section. The 350 cc YR5 model is fitted with a block tread rear tyre of 3.50 inch section. All models employ steel rims in conjunction with cast aluminium alloy hubs. Each wheel has, as standard, 8 inch diameter internal expanding brakes. In the case of the front wheel only, a twin leading shoe brake is used.
2 Both wheels are quickly detachable. The rear wheel can be removed from the frame without disturbing either the rear wheel sprocket or the final drive chain.

2 Front wheel: examination and renovation

1 Place the machine on the centre stand so that the front wheel is raised clear of the ground. Spin the wheel and check the rim alignment. Small irregularities can be corrected by tightening the spokes in the affected area, although a certain amount of practice is necessary to prevent over-correction. Any flats in the wheel rim should be evident at the same time. These are more difficult to remove and in most cases it will be necessary to have the wheel rebuilt on a new rim. Apart from the effect on stability, a flat will expose the tyre bead and walls to greater risk of damage.
2 Check for loose or broken spokes. Tapping the spokes is the best guide to tension. A loose spoke will produce a quite different sound and should be tightened by turning the nipple in an anti-clockwise direction. Always re-check for run-out by spinning the wheel again. If the spokes have to be tightened an excessive amount, it is advisable to remove the tyre and tube by the procedure detailed in Section 17 of this Chapter; this is so that the protruding ends of the spokes can be ground off, to prevent them from chafing the inner tube and causing punctures.

3 Front brake assembly: examination, renovation and re-assembly

1 The front brake assembly complete with brake plate can be withdrawn from the front wheel hub after the wheel spindle has been pulled out and the wheel removed from the forks. Refer to Chapter 4, Sections 2.6 and 2.7 for the correct procedure.
2 Examine the condition of the brake linings. If they are wearing thin or unevenly, the brake shoes should be renewed. The linings are bonded to the brake shoes and cannot be supplied separately.
3 To remove the brake shoes, turn the brake operating lever so that the brake is in the full-on position. Pull the brake shoes apart to free them from their operating cams and from the pivot on which the fixed ends bear. Then pull them upward, in a 'V' formation so that they can be lifted away together with the return springs. When they are well clear of the brake plate, the return springs can be detached.
4 Before replacing the brake shoes, check that both brake operating cams are working smoothly and not binding in their pivots. The cams can be removed for greasing by detaching the operating arm from the end of each shaft. The arms have splines which engage with similar splines on the end of each shaft; mark both the operating arm and the splined shaft before the arm is detached, to aid correct relocation. Do not alter the setting of the rod that joins the two operating arms, otherwise the brake setting will require adjustment after reassembly.
5 Check the inner surface of the brake drum, on which the brake shoes bear. The surface should be free from score marks indentations, otherwise reduced braking efficiency will be inevitable. Remove all traces of brake lining dust and wipe with a rag soaked in petrol, to remove all traces of grease and oil.

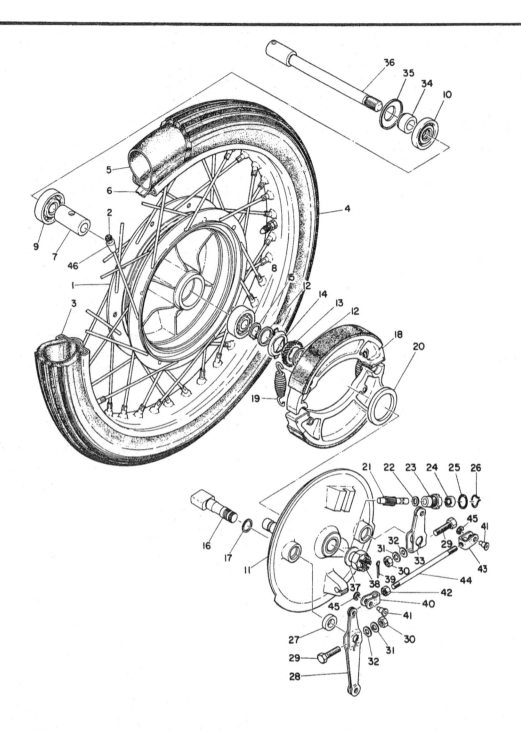

FIG. 5.1. FRONT WHEEL — COMPONENT PARTS

1 Front hub	13 Speedometer drive gear	25 'O' ring	37 Plain washer
2 Spoke set	14 Speedometer drive clutch	26 Locating clip	38 Nut
3 Rim	15 Circlip	27 Brake cam seal	39 Split pin
4 Tyre	16 Brake operating cam	28 Brake operating arm - long	40 Adjuster rod shackle
5 Tube	17 Shim for operating cam	29 Bolt	41 Clevis pin
6 Rim tape	18 Brake shoe complete	30 Nut	42 Nut
7 Bearing spacer	19 Brake shoe return spring	31 Spring washer	43 Adjuster rod shackle
8 Bearing	20 Oil seal	32 Plain washer	44 Adjuster rod
9 Bearing	21 Speedometer drive worm	33 Brake operating arm - short	45 Circlip
10 Oil seal	22 Thrust washer	34 Wheel spindle distance piece	46 Wheel balance weights -
11 Brake plate	23 Bush	35 Hub dust cover	number as required
12 Thrust washer	24 Oil seal	36 Wheel spindle	

6 To reassemble the brake shoes on the brake plate, fit the return springs and pull the shoes apart whilst holding them in the form of a 'V', facing upwards. If they are now located with the brake operating cams and fixed pivots, they can be pushed back into position by pressing downward. Do not use force, or there is risk of distorting the shoes.

4 Wheel bearings: examination and replacement

1 Access is available to the wheel bearings when the brake plate has been removed. The left-hand bearing is exposed when the brake plate is lifted after the oil seal, located in front, has been prised out of position.

2 Lay the wheel on the ground with the brake drum facing upward and with a special tool, in the form of a rod with a curved end, insert the curved end into the hole in the centre of the spacer separating the two wheel bearings. If the other end of the special tool is hit with a hammer, the right-hand bearing, bearing flange washer, and bearing spacer will be expelled from the hub.

3 Invert the wheel and drive out the left-hand bearing by inserting a drift of the appropriate size, through the hub. During the removal of either bearing it may be necessary to support the wheel across an open-ended box so that there is sufficient clearance for the bearing to be displaced completely from the hub.

4 Remove all the old grease from the hub and bearings, giving the latter a final wash in petrol. Check the bearings for signs of play or roughness when they are turned. If there is any doubt about the condition of a bearing, it should be renewed.

5 Before replacing the bearings, first pack the hub with new grease. Then drive the bearings back into position, not forgetting the distance piece that separates them. Fit replacement oil seals and any dust covers or spacers that were also displaced during the original dismantling operation.

5 Front wheel: reassembly and replacement

1 Place the front brake plate and brake assembly in the brake drum and align the wheel in the forks so that the cast-in slot of the brake plate engages with the abutment of the left-hand fork leg. This acts as the anchorage for the front brake and must be positioned correctly.

2 Align the wheel so that the front wheel spindle can be inserted from the right. Push the spindle home through the left-hand fork leg until the head of the spindle is flush with the right-hand fork leg. Replace the washer and castellated nut on the end of the fork spindle and tighten the nut. Replace the split pin that passes through the end of the nut and the spindle. Re-tighten the split clamp at the extreme end of the right-hand fork leg.

3 Spin the wheel to ensure that it moves freely, then attach the front brake cable and the speedometer drive cable, the latter being retained by a wire clip. Make sure that both cables pass through the cable retaining loop of the mudguard, to ensure that they cannot come into contact with either the tyre or wheel. Check that the brake functions correctly, especially if the brake operating arms have been removed and replaced. If necessary, readjust the brake by following the procedure described in Section 7 of this Chapter, then spin the wheel again as a final check.

6 Rear wheel: examination, removal and renovation

1 Place the machine on the centre stand, so that the rear wheel is raised clear of the ground. Check for rim alignment, damage to the rim and for loose or broken spokes, as described in Section 2 of this Chapter.

2 To remove the rear wheel, use the procedure recommended in Section 10.3 of Chapter 4, after detaching the final drive chain. If there is no necessity to disturb either the rear sprocket

3.1 Brake assembly is free after removal of front wheel

6.3a Remove distance piece after spindle has been withdrawn

6.3b Pull wheel sideways to disengage cush drive

or the chain, an amended procedure can be adopted, as follows:
Detach the rear brake torque arm from the brake plate by
removing the split pin through the torque arm bolt, then
remove the nut and bolt so that the torque arm is free from
the brake plate. Remove the rear brake rod by unscrewing the
adjuster, then pull the rod clear of the operating arm. Remove
the trunnion from the operating arm and the spring from the
brake rod, so that they are not mislaid.

3   Remove the split pin passing through the end of the rear
wheel spindle and the castellated nut. Then unscrew and remove
the castellated nut and withdraw the wheel spindle from the
right-hand side of the machine. If the distance piece between the
frame end and the brake plate is removed, the rear wheel can be
pulled sideways to disengage it from the cush drive of the rear
sprocket and then pulled clear from the rear of the machine. It
will be necessary to incline the machine to the left, so that there
is sufficient clearance for the wheel to be drawn from the
mudguard.

4   The rear brake plate and brake assembly can be withdrawn
from the right-hand side of the wheel hub.

5   The rear wheel bearings are a drive fit in the hub, separated
by a spacer, using a general arrangement similar to that of the
front wheel. Use a similar technique for removing, greasing and
replacing the bearings, noting that there is no oil seal in front of
the left-hand bearing.

7   Rear brake assembly: examination, renovation and re-assembly

1   The rear brake assembly complete with brake assembly can
be withdrawn from the rear wheel after the wheel spindle has
been pulled out, the distance piece removed and the wheel
pulled clear of the rear forks. The preceding Section describes a
simplified method of wheel removal if there is no necessity to
disturb either the rear wheel sprocket or the final drive chain.

2   If it is necessary to dismantle the rear brake assembly, follow
the procedure described in Section 3 of this Chapter. Note that
the rear brake is of the single leading shoe type and therefore
differs slightly in construction, having only one operating arm
for both brake shoes.

8   Adjusting the twin leading shoe front brake

1   If the front brake adjustment is correct, there should be a
clearance of not less than 20-30 mm (0.8 - 1.2 inches) between
the brake lever and the twist grip when the brake is applied fully.
There should also be from 5-8 mm (3/16 - 5/16 inch) free play in
the cable before the brake commences to operate. This clearance
is measured at the handlebar lever and is set by varying the cable
adjuster.

2   Adjustment is effected by turning the adjuster fitted to the
end of the handlebar lever clockwise to increase the clearance
and anti-clockwise to decrease the clearance. If the adjuster on
the front brake plate is used as an alternative, similar directions
apply in terms of rotation.

3   The threaded operating rod connecting the two brake
operating arms should not require attention unless the setting has
been disturbed. It is imperative that the leading edge of each
brake shoe contacts the brake drum simultaneously, if maximum
braking efficiency is to be achieved. Check by detaching the
clevis pin from the eye of one end of the threaded rod, after
removing the split pin, so that the brake operating arms can be
applied independently. Operate each arm in turn and note when
the brake shoe commences to touch the brake drum. Make a
mark to show the exact position of each operating arm when this
initial contact is made. Replace the clevis pin and check that the
marks correspond when the brake is applied in similar fashion. If
they do not, remove the clevis pin again and use the adjuster to
either decrease or increase the length of the rod joining the
two operating arms until the marks correspond. Replace
the clevis pin and do not omit the split pin that retains it in

7.2 Lift and pull brake shoes to separate from brake plate

8.3 Threaded rod links both brake operating arms

11.1 Remove large nut to detach sprocket and cush drive plate

position. Make sure the locknut on the rod that connects the brake operating arm is tightened. Re-check the brake lever adjustment, before the machine is used on the road.

4 A rough visual check can be made by checking whether the brake operating arms are exactly parallel to one another. If they are, the brake adjustment must be near-correct.

5 It is equally important to check that the brake pull-off action is good, otherwise there is risk of loss of power through binding brakes, and reduced braking efficiency as the linings wear or heat up. Sluggish action is usually due to a poorly lubricated brake cable, weak brake shoe return springs, or a tendency for the brake operating cams to bind in their bushes as a result of under-lubrication.

9 Adjusting the rear brake

1 If the adjustment of the rear brake is correct, the rear brake pedal will have about 25 mm (1 inch) free play before the brake commences to operate. Before the amount of travel is adjusted, the brake pedal should be positioned so that it is in the best position for quick operation. Because rider requirements vary, the pedal is retained to the shaft by splines, giving many variations of operating angle. A simple pinch-bolt arrangement holds the pedal on these splines, when the correct operating angle has been selected.

2 The length of travel is controlled by the adjuster at the end of the brake operating rod, close to the brake operating arm. If the nut is turned clockwise, the amount of travel is reduced and vice-versa. Always check that the brake is not binding after adjustments have been made.

3 Note that it may be necessary to re-adjust the height of the stop lamp switch if the pedal height has been changed to any marked extent. The switch located immediately below the right-hand side cover that carries the capacity symbol of the model. The body of the switch is threaded, so that it can be raised or lowered, after the lock nuts have been slackened. If the stop lamp lights too soon, the switch should be lowered, and vice-versa.

10 Cush drive assembly: examination and renovation

1 The cush drive assembly is contained within the left-hand side of the rear wheel hub. It comprises a set of synthetic rubber buffers, housed within a series of vanes cast in the hub shell. A plate attached to the centre of the rear wheel sprocket has four cast-in dogs which engage with slots in these rubbers, when the wheel is replaced in the frame. The drive to the rear wheel is transmitted via these rubbers, which cushion any surges or roughness in the drive which would otherwise convey the impression of harshness.

2 Examine the rubbers periodically for signs of damage or general deterioration. Renew and fit the rubbers as a set if there is any doubt about their condition; there is no difficulty in removing or replacing them as they are not under compression when the drive plate is attached.

11 Rear wheel sprocket: removal, examination and replacement

1 The rear wheel sprocket assembly can be removed as a separate unit after the rear wheel has been separated and detached from the frame as described in Sections 6.2 and 6.3 of this Chapter. Alternatively, it can be removed still attached to the rear wheel if the final drive chain is detached and the procedure described in Chapter 4.10, paragraph 3, is followed. In the wheel is free from the frame.

2 Check the condition of the sprocket teeth. If they are hooked, chipped or badly worn, the sprocket must be renewed. It is secured to the cush drive plate by four bolts and lock washers.

3 It is considered bad practice to renew one sprocket on its

11.2 Sprocket is retained by four bolts and lock washers

11.4a When circlip is removed, sprocket shaft ...

11.4b ... can be removed from cush drive plate

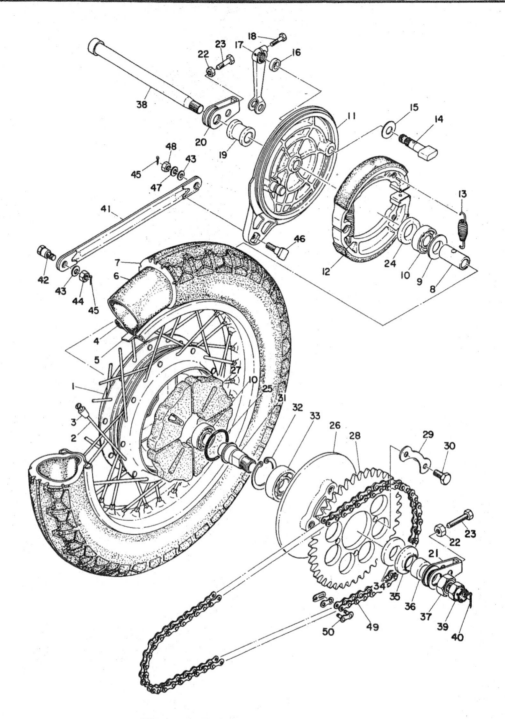

FIG. 5.2. REAR WHEEL — COMPONENT PARTS

1 Rear hub	13 Brake shoe return spring	25 'O' ring	38 Wheel spindle
2 Spoke set	14 Brake operating cam	26 Shock absorber hub	39 Wheel spindle nut
3 Wheel balance weights - number as required	15 Shim for brake operating cam	27 Shock absorber rubbers	40 Split pin
		28 Rear wheel sprocket	41 Rear brake torque arm
4 Rim	16 Cam seal	29 Lock washer	42 Bolt
5 Rim tape	17 Brake operating arm	30 Bolt	43 Plain washer
6 **Tube**	18 Bolt	31 Sprocket shaft	44 Nut
7 Tyre	19 Wheel spindle spacer	32 Circlip	45 Split pin
8 Bearing spacer	20 Chain adjuster - left-hand	33 Sprocket bearing	46 Bolt
9 Spacer flange	21 Chain adjuster - right-hand	34 Oil seal	47 Spring washer
10 Bearing	22 Nut	35 Dust cover	48 Nut
11 Brake plate	23 Chain adjuster screw	36 Sprocket shaft collar	49 Final drive chain
12 Brake shoe complete	24 Oil seal	37 Sprocket shaft nut	50 Spring link

Tyre changing sequence - tubed tyres

 A Deflate tyre. After pushing tyre beads away from rim flanges push tyre bead into well of rim at point opposite valve. Insert tyre lever adjacent to valve and work bead over edge of rim.

B Use two levers to work bead over edge of rim. Note use of rim protectors

 C Remove inner tube from tyre

 D When first bead is clear, remove tyre as shown

E When fitting, partially inflate inner tube and insert in tyre

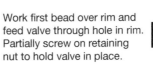 **F** Work first bead over rim and feed valve through hole in rim. Partially screw on retaining nut to hold valve in place.

 G Check that inner tube is positioned correctly and work second bead over rim using tyre levers. Start at a point opposite valve.

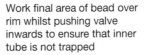 **H** Work final area of bead over rim whilst pushing valve inwards to ensure that inner tube is not trapped

12.3 Fork ends are marked to aid wheel alignment

12.7 Closed end of spring clip should face direction of chain travel

own. The final drive sprocket should always be renewed as a pair and a new chain fitted, otherwise rapid wear will necessitate even earlier renewal on the next occasion.

and a new chain fitted, otherwise rapid wear will necessitate even earlier replacement on the next occasion.

4   An additional bearing is located within the cush drive plate, which supports the sprocket shaft into which the rear wheel spindle fits. In common with the wheel bearings, this bearing is a journal ball and when wear occurs, the sprocket will give the appearance of being loose on its mounting bolts. The bearing is a tight push fit on the sprocket shaft and is preceded by an oil seal that excludes road grit and water. A circlip retains the assembly within the cush drive plate.

5   Remove the oil seal and bearing and wash out the latter to remove all traces of the old grease. If the bearing has any play or runs roughly, it must be renewed.

6   If the bearing has not been renewed it should be repacked with grease and pushed back on the sprocket shaft, followed by the oil seal. Replace the rear wheel assembly by reversing whichever method was adopted for its removal.

12 Final drive chain: examination and lubrication

1   The final drive chain is fully exposed, with only a light chainguard over the top run. Periodically the tension will need to be adjusted, to compensate for wear. This is accomplished by placing the machine on the centre stand and slackening the two wheel nuts on the left-hand side of the rear wheel so that the wheel can be drawn backward by means of the drawbolt adjusters in the fork ends. The rear brake torque arm bolt must also be slackened during this operation.

2   The chain is in correct tension if there is approximately 20 mm (¾ inch) slack in the middle of the lower run. Always check when the chain is at its tightest point as a chain rarely wears evenly during service.

3   Always adjust the drawbolts an equal amount in order to preserve wheel alignment. The fork ends are clearly marked with a series of horizontal lines above the adjusters, to provide a simple, visual check. If desired, wheel alignment can be checked by running a plank of wood parallel to the machine, so that it touches the side of the rear tyre. If wheel alignment is correct, the plank will be equidistant from each side of the front wheel tyre, when tested on both sides of the rear wheel. It will not touch the front wheel tyre because this tyre is of smaller cross section. See accompanying diagram.

4   Do not run the chain overtight to compensate for uneven wear. A tight chain will place undue stress on the gearbox and rear wheel bearings, leading to their early failure. It will also absorb a surprising amount of power.

5   After a period of running, the chain will require lubrication. Lack of oil will greatly accelerate the rate of wear of both the chain and the sprockets and will lead to harsh transmission. The application of engine oil will act as a temporary expedient, but it is preferable to remove the chain and clean it in a paraffin bath before it is immersed in a molten lubricant such as "Linklyfe" or "Chainguard". These lubricants achieve better penetration of the chain links and rollers and are less likely to be thrown off when the chain is in motion.

6   To check whether the chain is due for replacement, lay it lengthwise in a straight line and compress it endwise so that all the play is taken up. Anchor one end and measure the length. Now pull the chain with one end anchored firmly, so that the chain is fully extended by the amount of play in the opposite direction. If there is a difference of more than ¼ inch per foot in the two measurements, the chain should be replaced in conjunction with the sprockets. Note that this check should be made AFTER the chain has been washed out, but BEFORE any lubricant is applied, otherwise the lubricant may take up some of the play.

7   When replacing the chain, make sure that the spring link is seated correctly, with the closed end facing the direction of travel.

13 Chain and sprocket conversions

1   It is now possible to obtain a British-made chain and sprocket conversion kit which permits non-metric components to be substituted. The main advantage is the use of a heavier weight of chain and wider sprockets, reducing the rate of wear by a significant amount. It follows that the kit must be purchased as a whole, since metric and non-metric parts cannot be used in conjunction with one another.

2   Despite widespread belief to the contrary, there is rarely any advantage to be gained from varying sprocket sizes from those specified by the manufacturer. A larger gearbox sprocket by no means guarantees a higher maximum speed; usually it converts top gear into little more than an overdrive and may even lead to a situation where the machine is faster in the next lower gear.

14 Tyres: removal and replacement

1   At some time or other the need will arise to remove and replace the tyres, either as the result of a puncture or because a renewal is required to offset wear. To the inexperienced, tyre changing represents a formidable task yet if a few simple rules are observed and the technique learned, the whole operation is surprisingly simple.

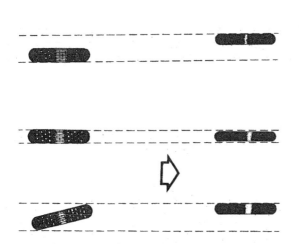

Fig. 5.3. Method of checking wheel alignment

2 To remove the tyre from the wheel, first detach the wheel from the machine by following the procedure in Chapter 4, Sections 2.6 and 2.7 or Sections 6.2 and 6.3 of this Chapter, depending on whether the front or the rear wheel is involved. Deflate the tyre by removing the valve insert and when it is fully deflated, push the bead of the tyre away from the wheel rim on both sides so that the bead enters the centre well of the rim. Remove the locking cap and push the tyre valve into the tyre itself.

3 Insert a tyre level close to the valve and lever the edge of the tyre over the outside of the wheel rim. Very little force should be necessary; if resistance is encountered it is probably due to the fact that the tyre beads have not entered the well of the wheel rim all the way round the tyre.

4 Once the tyre has been edged over the wheel rim, it is easy to work around the wheel rim so that the tyre is completely free on one side. At this stage, the inner tube can be removed.

5 Working from the other side of the wheel, ease the other edge of the tyre over the outside of the wheel rim furthest away. Continue to work around the rim until the tyre is free from the rim.

6 If a puncture has necessitated the removal of the tyre, reinflate the inner tube and immerse it in a bowl of water to trace the source of the leak. Mark its position and deflate the tube. Dry the tube and clean the area around the puncture with a petrol-soaked rag. When the surface has dried, apply rubber solution and allow this to dry before removing the backing from the patch and applying the patch to the surface.

7 It is best to use a patch of the self-vulcanising type, which will form a very permanent repair. Note that it may be necessary to remove a protective covering from the top surface of the patch, after it has sealed in position. Inner tubes made from synthetic rubber may require a special type of patch and adhesive, if a satisfactory bond is to be achieved.

8 Before replacing the tyre, check the inside to make sure the agent that caused the puncture is not trapped. Check also the outside of the tyre, particularly the tread area, to make sure nothing is trapped that may cause a further puncture.

9 If the inner tube has been patched on a number of past occasions, or if there is a tear or large hole, it is preferable to discard it and fit a new one. Sudden deflation may cause an accident, particularly if it occurs with the front wheel.

10 To replace the tyre, inflate the inner tube sufficiently for it to assume a circular shape but only just. Then push it into the tyre so that it is enclosed completely. Lay the tyre on the wheel at an angle and insert the valve through the rim tape and the hole in the wheel rim. Attach the locking cap on the first few threads, sufficient to hold the valve captive in its correct location.

11 Starting at the point furthest from the valve, push the tyre bead over the edge of the wheel rim until it is located in the central well. Continue to work around the tyre in this fashion until the whole of one side of the tyre is on the rim. It may be necessary to use a tyre lever during the final stages.

12 Make sure there is no pull on the tyre valve and again commencing with the area furthest from the valve, ease the other bead of the tyre over the edge of the rim. Finish with the area close ot the valve, pushing the valve up into the tyre until the locking cap touches the rim. This will ensure the inner tube is not trapped when the last section of the bead is edged over the rim with a tyre lever.

13 Check that the inner tube is not trapped at any point. Reinflate the inner tube, and check that the tyre is seating correctly around the wheel rim. There should be a thin rib moulded around the wall of the tyre on both sides, which should be equidistant from the wheel rim at all points. If the tyre is unevenly located on the rim, try bouncing the wheel when the tyre is at the recommended pressure. It is probable that one of the beads has not pulled clear of the centre well.

14 Always run the tyres at the recommended pressures and never under or over-inflate. The correct pressures for solo use are given in the Specifications section of this Chapter.

15 Tyre replacement is aided by dusting the side walls, particularly in the vicinity of the beads, with a liberal coating of french chalk. Washing-up liquid can also be used to good effect, but this has the disadvantage of causing the inner surfaces of the wheel rim to rust.

16 Never replace the inner tube and tyre without the rim tape in position. If this precaution is overlooked there is good chance of the ends of the spoke nipples chafing the inner tube and causing a crop of punctures.

17 Never fit a tyre that has a damaged tread or side walls. Apart from the legal aspects, there is a very great risk of a blow-out, which can have serious consequences on any two-wheel vehicle.

18 Tyre valves rarely give trouble, but it is always advisable to check whether the valve itself is leaking before removing the tyre. Do not forget to fit the dust cap, which forms an effective second seal.

Fault diagnosis overleaf

15 Fault diagnosis — Wheels, brakes and tyres

Sympton	Reason/s	Remedy
Handlebars oscillate at low speeds	Buckled front wheel Incorrectly fitted front tyre	Remove wheel for specialist attention. Check whether line around bead is equidistant from rim.
Forks 'hammer' at high speeds	Front wheel out of balance	Add weights until wheel will stop in any position.
Brakes grab, locking wheel	Ends of brakes shoes not chamfered	Remove brake shoes and chamfer ends.
Brakes feel spongy	Stretched brake operating cables, weak pull-off springs.	Replace cables and/or springs, after inspection.
Tyres wear more rapidly in middle of tread	Over inflation	Check pressures and run at recommended settings.
Tyres wear rapidly at outer edges of tread	Under-inflation	Ditto.

Chapter 6 Electrical system

Contents

Specifications

Battery

Type	Lead-acid
Make	Furukawa AYT2 – 12
Voltage	12 volts
Capacity	5.5 amp hrs

Alternator

Make	Mitsubishi
Type	AZ – 201ON
Voltage	12 volts
Output	238 watts

Regulator unit

Make	Mitsubishi
Type	RN – 2226J3
Open circuit adjustment	15.5 – 16.5 volts

Bulbs

Main headlamp	35/35W, pre-focus
Pilot lamp	3W bayonet fitting
Tail/stop lamp	8/23W offset pins bayonet fitting
Speedometer lamp	3W bayonet fitting
Tachometer lamp	3W bayonet fitting
Main beam indicating lamp	3W bayonet fitting
Flashing indicator lamp	3W bayonet fitting
Neutral indicator lamp	3W bayonet fitting
Flashing indicator lamps	8W each (4), bayonet fitting

All bulbs 12 volt rating.

1 General description

1 The 250 cc and 350 cc Yamaha twins are fitted with a 12 volt electrical system. The circuit comprises a crankshaft-driven alternator, the output of which is controlled by a voltage regulator linked with the stator coil windings. Because the output from the alternator is ac a rectifier is included in the circuit to convert current to dc in order to maintain the charge of the 12 volt, 5.5 amp hour battery.

2 Because the alternator has coils in the rotor assembly, brush gear is employed to pick up the current for the rotor from slip rings on the outer face of the rotor. No permanent magnets are employed in the construction of the alternator; it functions solely on the electro-magnetic principle.

2 Alternator: checking the output

1 As previously mentioned in Chapter 3.2, there is no satisfactory method of checking the output from the alternator without test equipment of the multi-meter type. If the performance of the alternator is in any way suspect, it should be checked by either a Yamaha repair specialist or auto-electrical mechanic.

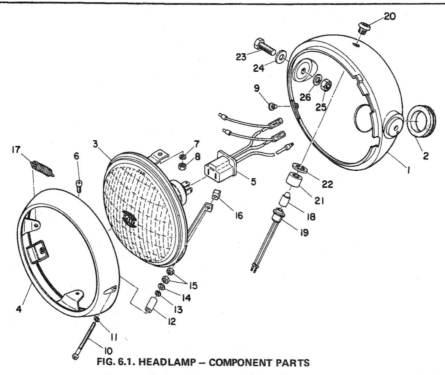

FIG. 6.1. HEADLAMP – COMPONENT PARTS

1	Headlamp shell	7	Spring washer
2	Grommet	8	Nut
3	Reflector unit	9	Rim retaining screw
4	Rim	10	Headlamp adjusting screw
5	Electrical socket	11	Washer
6	Screw	12	Spacer

13	Plain washer	20	Main beam warning lamp
14	Spacer	21	Warning lamp nut
15	Nut	22	Toothed washer
16	Adjuster nut	23	Bolt
17	Adjuster spring	24	Plain washer
18	Warning bulb (1.5 W)	25	Nut
19	Socket cord assembly	26	Spring washer

3 Voltage regulator: checking the controlling action

1 It is the function of the voltage regulator to pass a controlled amount of voltage to the rotor windings so that the output from the alternator is matched to the requirements of the electrical system. In effect, the voltage regulator can be regarded as a magnetic switch. A spring tensioned arm above an electro-magnet carries contacts, which will make or break according to the pull of the magnet. By this means it is possible to keep the voltage supplied by the alternator within the 12-15 volt range so that there are no surges or sudden overloads.

2 The voltage regulator is located at the base of the battery carrier, below the battery. Access is gained by raising the dual seat and removing the battery, then detaching the cover over the regulator unit which is retained by two small cross head screws.

3 To check whether the regulator is functioning correctly, start the engine then remove the red lead connecting the voltage regulator to the rectifier. Connect a voltmeter with a 0.20 volt dc range so that the positive lead replaces the connection previously made to the rectifier and attach the negative lead to the frame or some other convenient earthing point. Increase the engine speed to about 2,500 rpm and check the voltmeter reading. If the adjustment is correct the reading should be within the range 15.5 - 16.5 volts (off load).

4 If adjustment is needed, bend the adjusting arm A SMALL AMOUNT either upward or downward - upward to increase the charge rate or downward to effect a reduction.

5 A defective voltage regulator can cause either an abnormally high or an abnormally low output. Apart from re-setting the adjusting arm it is not practicable to repair a voltage regulator that malfunctions with the exception of cleaning the contacts. Replacement will be necessary if the regulator fails to respond to this treatment.

4 Battery: examination and maintenance

1 A Furukawa type AYT2-12 battery is fitted as standard. This battery is a lead-acid type and has a capacity of 5.6 amp hours.

2 The transparent plastic case of the battery permits the upper and lower levels of the electrolyte to be observed when the battery is lifted from its housing below the dual seat. Maintenance is normally limited to keeping the electrolyte level between the prescribed upper and lower limits and by making sure the vent pipe is not blocked. The lead plates and their separators can be seen through the transparent case, a further guide to the general condition of the battery.

3 Unless acid is spilt, as may occur if the machine falls over, the electrolyte should always be topped up with distilled water, to restore the correct level. If acid is spilt on any of the machine, it should be neutralised with an alkali such as washing soda and washed away with plenty of water, otherwise serious corrosion will occur. Top up with sulphuric acid of the correct specific gravity (1.260 - 1.280) only when spillage has occurred. Check that the vent pipe is well clear of the frame tubes or any of the other cycle parts, for obvious reasons.

5 Battery: charging procedure

1 The normal charging rate for the 5.5 amp hour battery is 0.5 amps. A more rapid charge, not exceeding 1 amp can be given in an emergency. The higher charge rate should, if possible, be avoided since it will shorten the working life of the battery.

2 Make sure that the battery charger connections are correct, red to positive and black to negative. It is preferable to remove the battery from the machine whilst it is being charged and to remove the vent plug from each cell. When the battery is re-connected to the machine, the black lead must be connected

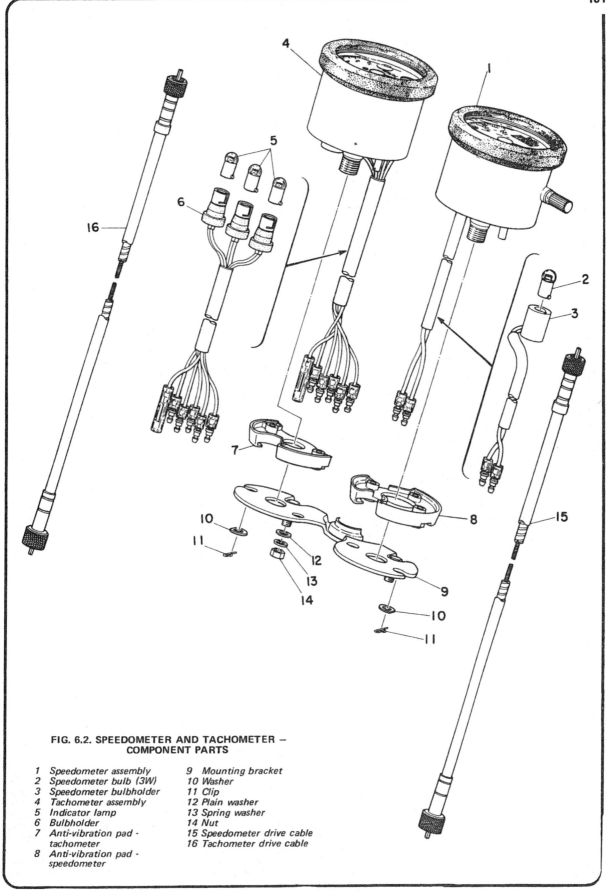

**FIG. 6.2. SPEEDOMETER AND TACHOMETER —
COMPONENT PARTS**

1 Speedometer assembly
2 Speedometer bulb (3W)
3 Speedometer bulbholder
4 Tachometer assembly
5 Indicator lamp
6 Bulbholder
7 Anti-vibration pad -
tachometer
8 Anti-vibration pad -
speedometer
9 Mounting bracket
10 Washer
11 Clip
12 Plain washer
13 Spring washer
14 Nut
15 Speedometer drive cable
16 Tachometer drive cable

4.2 Transparent case of battery makes electrolyte level inspection easy

5.1 Remove vent plugs when re-charging the battery

8.1 Small screw retains headlamp rim and reflector

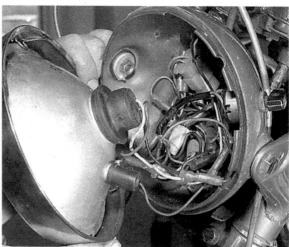

8.2a Bulb holders are attached to reflector by rubber sleeves

to the negative terminal and the red lead to positive. This is most important, as the machine has a negative earth system. If the terminals are inadvertently reversed, the electrical system will be damaged permanently. The rectifier will be destroyed by a reversal of the current flow.

6 Rectifier: general description

1   This is a full wave rectifier composed of nine silicon diodes. The diodes permit a one-way flow of electrical current and therefore convert the alternating current output from the alternator into direct current, which can be used for battery charging.

2   In the event of failure of the battery to maintain a fully-charged condition, it is possible that the rectifier is malfunctioning. Unfortunately there is no easy way of checking without the appropriate electronic test equipment. Provided the electrical connections have not been inadvertently transposed at the battery, a check by substitution of the correct replacement is the only practicable method of verification.

7 Fuse: location and replacement

1   A fuse is incorporated in the electrical system to give protection from sudden overload, which may occur during a short circuit. The fuse is contained within a plastic fuseholder that forms part of the electrical wiring. It is located close to the positive terminal of the battery, under the dual seat. The fuse is rated at 20 amps; always carry a spare fuse for use in an emergency.

2   Before replacing a fuse that has blown, check that no obvious short circuit has occurred, otherwise the replacement fuse will blow immediately it is inserted. It is always wise to check the electrical circuit thoroughly, to trace the fault and eliminate it.

3   When a fuse blows while the machine is running and no spare is available, a 'get you home' remedy is to remove the blown fuse and wrap it in silver paper before replacing it in the fuseholder. The silver paper will restore the electrical continuity by bridging the broken fuse wire. This expedient should NEVER be used if there is evidence of a short circuit or other major electrical fault, otherwise more serious damage will be caused. Replace the 'doctored' fuse at the earliest possible opportunity, to restore full circuit protection.

8.2b Main headlamp bulb is of pre-focus, double filament type

8.4 Pilot lamp bulbholder has bayonet fitting

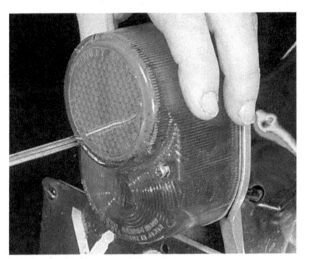

10.1 Two screws retain lens of rear lamp

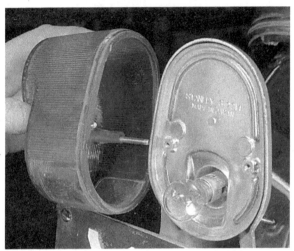

10.2 Bulb has staggered pins to prevent reversal of contacts

8 Headlamp: replacing bulbs and adjusting beam height

1 To remove the headlamp rim, detach the small screw on the right-hand underside of the headlamp shell. The rim can then be prised off, complete with the reflector unit.

2 The main bulb is a twin filament type, to give a dipped beam facility. The bulb holder is attached to the back of the reflector by a rubber sleeve, which fits around a flange in the reflector and the flange of the bulbholder. An indentation in the bulbholder orifice and a projection on the bulbholder ensures the bulb is always replaced in the same position so that the focus is unaltered.

3 It is not necessary to re-focus the headlamp when a new bulb is fitted. Apart from the just-mentioned method of location, the bulbs used are of the pre-focus type, built to a precise specification. To release the bulb from the bulb holder, twist and lift away.

4 The pilot lamp bulb, like the bulb itself, has a bayonet fitting. It is protected by a rubber sleeve. Remove the bulb holder first, then the bulb.

5 The main headlamp bulb is rated at 35/35w, 12 volts and the pilot lamp bulb at 3w, 12 volts. Variations in the wattage may occur according to the country or state for which the machine is supplied. In the UK, the pilot bulb a statutory minimum rating of 6w.

6 Beam alignment is adjusted by tilting the headlamp after the two retaining bolts have been slackened and then retightening them after the correct beam height is obtained, without moving the setting.

7 To obtain the correct beam height, place the machine on level ground facing a wall 25 yards distant, with the rider seated normally. The height of the beam centre should be equal to that of the height of the centre of the headlamp from the ground, when the dip switch is in the main beam position. Furthermore, the concentrated area of light should be centrally disposed. Adjustments in either direction are made by rearranging the angle of the headlamp, as described in the preceding paragraph. Note that a different beam setting will be needed when a pillion passenger is carried. If a pillion passenger is carried regularly, the passenger should be seated in addition to the rider when the beam setting adjustment is made.

9 The above instructions for beam setting relate to the requirements of the United Kingdom's transport lighting regulations. Other settings may be required in countries other than the UK.

9 Handlebar switches: function and replacement

1 The dipswitch froms part of the left-hand dummy twist grip which contains the horn button, flashing indicator lamp switch,

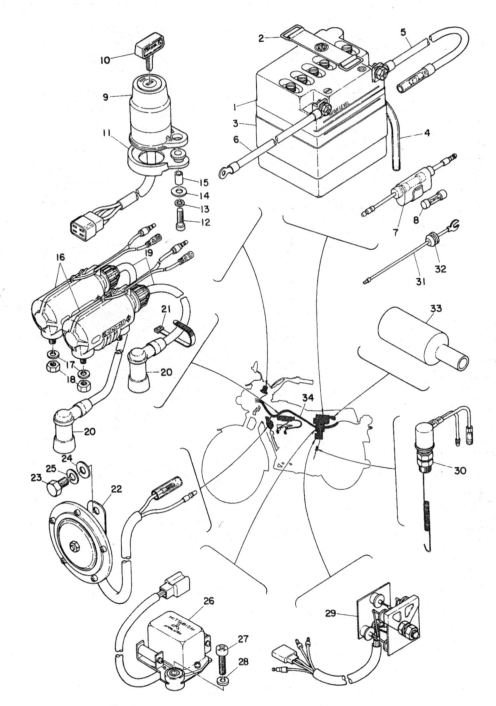

FIG. 6.3. ELECTRICAL EQUIPMENT — COMPONENT PARTS

1 Battery	13 Spring washer	25 Spring washer
2 Battery retaining strap	14 Washer	26 Voltage regulator
3 Battery holder	15 Switch collar	27 Screw
4 Vent pipe	16 Ignition coil	28 Spring washer
5 Positive lead	17 Spring washer	29 Rectifier
6 Negative lead	18 Nut	30 Stop lamp switch
7 Fuse holder	19 High tension lead	31 Lead wire
8 Fuse link	20 Plug cap	32 Grommet
9 Ignition and lighting switch	21 Cable clip	33 Connector cover
10 Switch key	22 Horn	34 Wiring harness
11 Switch mounting	23 Bolt	
12 Screw	24 Plain washer	

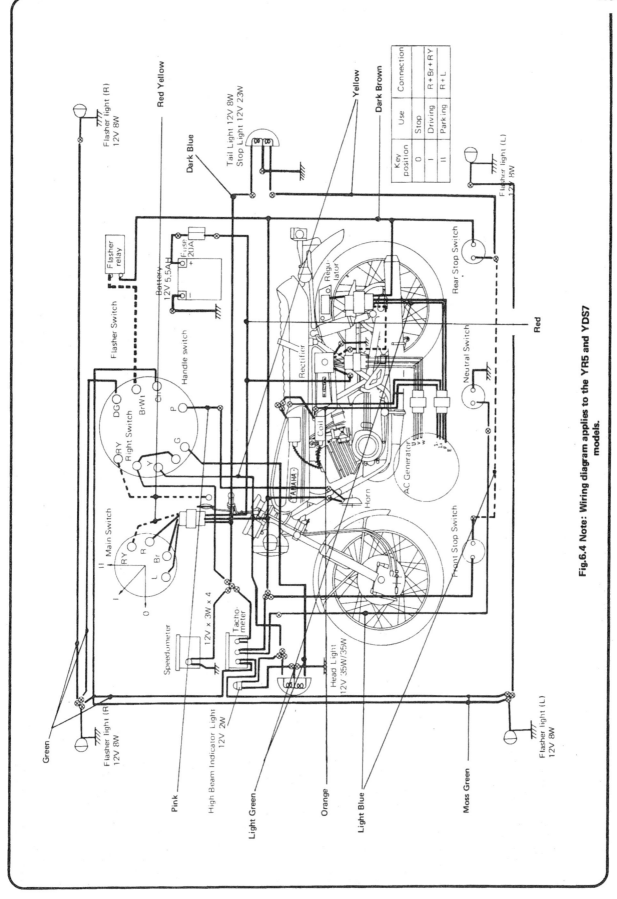

Fig.6.4 Note: Wiring diagram applies to the YR5 and YDS7 models.

Key position	Use	Connection
0	Stop	
I	Driving	R + Br + RY
II	Parking	R + L

and headlamp flasher.

2   In the event of failure of any of these switches, the switch assembly must be replaced as a complete unit since it is not practicable to effect a permanent repair.

3   A stop lamp switch, actuated by the front brake levers, is fitted to machines where traffic regulations necessitate stop lamp indication when either brake is applied.

10 Stop and tail lamp: replacing the bulb

1   The tail lamp is fitted with a twin filament bulb of 12 volt, 8/23W rating, to illuminate the rear number plate and rear of the machine, and to give visual warning when the rear brake is applied. To gain access to the bulb, remove the plastic lens cover, which is retained by two long screws. Check that the gasket between the lens cover and the main body of the lamp is in good condition.

2   The bulb has a bayonet fitting and has staggered pins to prevent the bulb contacts from being reversed.

3   If the tail lamp keeps blowing, suspect either vibration of the rear mudguard, or more probably, an intermittent earth connection.

11 Flashing indicator lamps

1   The forward facing indicator lamps are connected to 'stalks' that replace the bolts on which the headlamp shell would normally be mounted. The stalks are hollow and have threaded ends so that they can be locked in position from the inside of the headlamp shell or to the fork lugs that carry the headlamp. The rear facing lamps are mounted on similar, shorter stalks, at a point immediately to the rear of the dual seat.

2   In each case, access to the bulb is gained by removing the plastic lens cover, which is retained by two screws. Bayonet fitting bulbs of the single filement type are used, each with a 12 volt, 8w rating.

12 Flasher unit: location and replacement

1   The flasher relay unit is located either under the dual seat or behind the right-hand side cover that carries the capacity symbol of the model. It is retained by a single bolt passing through a built-in clip.

2   If the flasher unit is functioning correctly, a series of audible clicks will be heard when the indicator lamps are in action. If the unit mulfunctions and all the bulbs are in working order, the usual sympton is one initial flash before the unit goes dead; It will be necessary to replace the unit complete if the fault cannot be attributed to any other cause.

3   Take great care when handling a flasher unit because it is easily damaged if dropped.

13 Tachometer head: replacement of bulbs

1   The tachometer head houses three bulbs, one for illuminating the dial when the lights are in use, one for indicating when the flashing indicators are in use (yellow) and one to show when the gearbox is in neutral (green). All three bulbs are rated at 12 volts, 3 w and are of the bayonet fitting type.

2   The bulb holders are a push fit into the base of the tachometer head, where they are retained by their outer moulded rubber sleeves.

14 Speedometer head: replacement of bulb

1   The speedometer illuminating bulb is rated at 12 volts, 3 w, bayonet type. The bulb holder is a push fit into the base of the speedometer head and is retained in a similar manner to that employed for the tachometer bulbs.

15 Headlamp beam indicator lamp: replacement of bulb

1   An indicator bulb (red) is contained in the top of the headlamp shell which is illuminated when the headlamp is on main beam. The bulb is of 12 volt, 3 w rating bayonet type.

2   The bulbholder is located within the headlamp shell, making it necessary to remove the headlamp rim and reflector to gain access. It is a push fit into the indicator housing, where it is retained by a rubber moulding.

16 Horn: location and examination

1   The horn is suspended from a flexible steel strip bolted immediately below the steering head, between the duplex down tubes of the frame. The flexible strip isolates the horn from vibration.

2   The horn has no external means of adjustment. If it malfunctions, it must be renewed; it is a statutory requirement that the machine must be fitted with a horn in working order).

17 Wiring: layout and examination

1   The wiring harness is colour-coded and will correspond with the accompanying wiring diagram. Where socket connectors are used, they are designed so that reconnection can be made in the correct position only.

2   Visual inspection will show whether there are any breaks or frayed outer coverings which will give rise to short circuits. Another source of trouble may be the snap connectors and sockets, where the connector has not been pushed fully home in the outer housing.

3   Intermittent short circuits can often be traced to a chafed wire that passes through or is close to a metal component such as a frame member. Avoid tight bends in the lead or situations where a lead can become trapped between castings.

18 Ignition and lighting switch

1   The ignition and lighting switch is combined in one unit, bolted to the top fork yoke from the underside. It is operated by a key, which cannot be removed when the ignition is switched on.

2   The number stamped on the key will match the number of the steering head lock and that of the lock in the petrol filler cap. A replacement key can be obtained if the number is quoted; if either of the locks or the ignition switch is changed, additional keys will be required.

3   It is not practicable to repair the ignition switch if it malfunctions. It should be renewed with a new switch and key to suit.

19 Fault diagnosis - Electrical system

Symptom	Reason/s	Remedy
Complete electrical failure	Blown fuse	Check wiring and electrical components for short circuit before fitting new 15 amp fuse.
	Isolated battery	Check battery connections, also whether connections show signs of corrosion.
Dim lights, horn inoperative	Discharged battery	Recharge battery with battery charger and check whether alternator is giving correct output (electrical specialist).
Constantly 'blowing' bulbs	Vibration, poor earth connection	Check whether bulb holders are secured correctly. Check earth return or connections to frame.

Chapter 7 RD 250 and RD 350 models

Contents

Specifications

Clutch
No. of plates:

Plain ... ... Six

Inserted ... ... Seven

Gear ratios

	RD250 model		RD350 model	
First	2.571 : 1	2.571 : 1*	2.571 : 1	2.571 : 1*
Second	1.777 : 1	1.777 : 1	1.777 : 1	1.777 : 1
Third	1.318 : 1	1.318 : 1	1.318 : 1	1.318 : 1
Fourth	1.083 : 1	1.040 : 1	1.083 : 1	1.040 : 1
Fifth	0.961 : 1	0.888 : 1	0.961 : 1	0.888 : 1
Sixth	0.888 : 1	0.785 : 1	0.888 : 1	0.785 : 1
Primary drive reduction	3.238: 1		2.870 : 1	
Final drive reduction	2.666 : 1		2.668 : 1	

*Early RD models, with blanked off sixth gear (not USA)

Spark plugs
Type:

1973/74 RD250 and 350 models	NGK B8HS or ND W20FS-U
1975 RD250 and 350 models	NGK B8ES or ND W20ES-U
1976/77 RD250 models	NGK B7ES or ND W22ES-U
1978/79 RD250 models	NGK B9ES or ND W27ES-U

Reach:

1973/74 models ... ... ½ in

1975 on models ... ... ¾ in

Gap:

1973 - 1977 models ... ... 0.6 - 0.7 mm (0.024 - 0.028 in)

1978/79 models ... ... 0.7 - 0.8 mm (0.028 - 0.031 in)

1 General description

The RD250 and RD350 models were first introduced to the UK during March 1974. They are similar to the YDS7 and the YR5 models which they replaced and differ in only a few ways. Some modification has been made to the gear selector mechanism and arrangement of the gear ratios which now means that the previously 'blanked off' sixth gear is usable. The new RD models have an hydraulically operated disc brake which replaces the old internally expanding drum brake. Reed valves are now incorporated in the induction system which gives more positive control over the admission of the fuel/air mixture to the crankcase and thereby gives better performance and fuel consumption.

In 1976 the RD350 model was discontinued and the RD250 model was replaced by the RD250DX variant. At first glance the most noticeable change from the old RD250 to the new DX version is its almost completely new look. The tank has been altered and so have the dualseat, rear light, instrument covers, and front mudguard.

The 1978 version of the RD250DX model has had several further modifications. The new model has a CDI (capacitor discharge ignition) unit which replaces the previously employed contact breaker system. The kickstarter has been altered from the ratchet type to the bendix type and the silencer design has also been changed.

A noticeable difference found when sitting astride the 1978 model is the handlebars which are much narrower and slightly lower. This modification has been carried out to improve the riding position, which has also necessitated changing the position of the footrests.

As far as taking the engine apart is concerned there is little or no change in procedure. Any changes in procedure differing from that described in the preceding Chapters, are dealt with in this Chapter.

2 Clutch and six-speed gearbox

Clutch

1 The removal, examination and refitting information given in Chapter 1 can be followed, with the exception of that relating to the clutch plates. Fit the inserted and plain plates alternately, starting and finishing with an inserted plate. Note that a cushion ring should be installed between the clutch inner drum and first plain plate, and between the remaining plain plates.
2 On 1976-on models the plain plates have a pip on their periphery, and it is important to arrange the plain plates so that their pips are 60° from each other (see accompanying illustration). When fitting the pressure plate, align one of its three arrow marks with the corresponding mark on the clutch inner drum.

Six-speed gearbox

3 As mentioned earlier, the new gearbox contains an extra gear bringing it from a five-speed gearbox to a six-speed gearbox. Initially, the gear selector mechanism of the earlier RD model was arranged so that the sixth speed could not be selected and used, the gearbox to all intents and purposes functioning as though it were of the five-speed variety. At least one magazine feature showed how the gear selector mechanism could be modified so that the extra gear became usable, but Mitsui Machinery Sales (UK) Limited did not endorse this unauthorised modification because they claimed it would give rise to certain difficulties. For this reason, modification details are not included in this manual. Later models now incorporate a modified selector mechanism devised by the manufacturer together with some rearrangement of the internal gear ratios. On these models, sixth gear can now be selected positively, converting the gearbox to full six-speed operation.

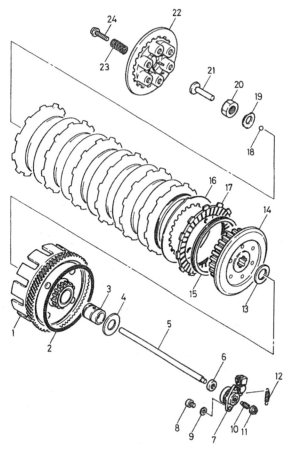

Fig. 7.1a Clutch component parts

1	Clutch outer drum	13	Thrust washer
2	O-ring	14	Clutch inner drum
3	Spacer	15	Cushion ring – 7 off
4	Thrust washer	16	Plain clutch plate – 6 off
5	Push rod	17	Inserted clutch plate – 7 off
6	Oil seal	18	Steel ball
7	Actuating mechanism	19	Belville washer
8	Screw – 2 off	20	Nut
9	Washer – 2 off	21	Push rod mushroom
10	Adjusting screw	22	Pressure plate
11	Locknut	23	Spring – 6 off
12	Spring	24	Screw – 6 off

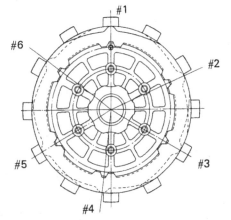

Fig. 7.1b. Clutch plain plate positioning – 1976-on models

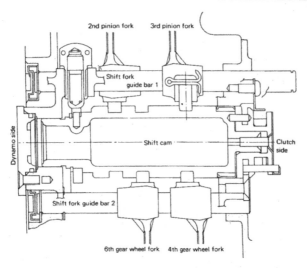

Fig. 7.1c. Arrangement of the gear selector forks – 6-speed models

3 Reed valve induction system: mode of operation and examination

1 Of the various systems of controlling the induction cycle of a two-stroke engine, Yamaha has chosen to adopt the reed valve, a device which permits precise control of the incoming mixture, allowing more favourable port timing to give improved torque and power outputs. The reed valve assembly comprises a wedge-shaped die-cast aluminium alloy valve case mounted in the inlet tract. The valve case has rectangular ports which are closed off by flexible stainless steel reeds. The reeds seal against a heat and oil resistant synthetic rubber gasket which is bonded to the valve case. A specially shaped valve stopper, made from cold rolled stainless steel plate, controls the extent of movement of the valve reeds.

2 As the piston ascends in the cylinder, a partial vacuum is formed beneath the cylinder in the crankcase. This allows atmospheric pressure to force the valves open, and a fresh charge of petrol/air mixture flows past the valve and into the crankcase. As the pressure differential becomes equalised, the valves close, and the incoming charge is then trapped. The charge of mixture in the cylinder is by this time fully compressed, and ignition takes place driving the piston downwards. The descending piston eventually uncovers the exhaust port, and the hot exhaust gases, still under a certain amount of pressure, are discharged into the exhaust system. At this stage, the reed valve, in conjunction with the 7th, or auxiliary scavenging port, performs a secondary function; as the hot exhaust gases rush out of the exhaust port, a momentary depression is created in the cylinder, this allows the valve to open once more, but this time the incoming mixture enters directly into the cylinder via the 7th port and completes the expulsion of the now inert burnt gases. This ensures that the cylinder is filled with the maximum possible combustion mixture. The charge of combustion mixture which has been compressed in the crankcase is released into the cylinder via the transfer ports, and the piston again ascends to close the various ports and begin compression. The reed valves open once more as another partial vacuum is created in the crankcase, and the cycle of induction thus repeats. It will be noted that no direct mechanical operation of the valve takes place, the pressure differential being the sole controlling factor.

4 Carburettors

The 1978 RD250DX model has a synchronization port which makes setting the carburettors much easier.

The actual settings are as follows:

Type	1977 model	1978 model
	VM28SS	VM28SS
No.	IS200	2R800
Main jet	115	115
Air jet	1.0	0.8
Jet needle	5L3–2	5L1–4
Needle jet	0.8	0.6
Throttle valve cutaway	2.5	2.0
Air screw	1¼ turns out	1¼ turns out
Starter jet	70	70

5 Oil pump adjustment: later models

1 Several modifications have been made which require different working procedures when adjusting the oil pump settings. These are described below but if there is any doubt at all about which, if any, apply to your machine, take the complete machine to a good Yamaha dealer for expert advice. At the very least make an accurate drawing of the oil pump pulley, showing all marks etc, and take this with the complete engine and frame numbers to the dealer so that he can determine the correct procedure for your model.

2 Before the pump is adjusted the carburettors must be synchronised and the throttle cable adjusted. On later models, each carburettor has a hexagon-headed plug screwed into its upper body, both plugs facing to the right-hand side of the machine. Remove these, open fully the throttle and look for a circular indentation drilled in the surface of each throttle slide; use the adjuster on each carburettor top to ensure that the lower edge of each indentation aligns exactly with the lower edge of its respective aperture. Secure the adjuster locknuts, open and close fully the throttle once or twice to settle the cables, then recheck the setting. Secure the adjuster locknuts, slide down the rubber adjuster covers and refit the inspection plugs. Using the adjuster beneath the twistgrip, adjust the throttle cable until there is 3-7 mm ($\frac{1}{8}$-$\frac{1}{4}$ in approx) free play, measured in terms of twistgrip rotation.

3 When checking the oil pump itself, note that on later models the white plastic pinion has been removed. If this is found to be the case it will be necessary to start the engine and to allow it to idle while watching closely the adjuster plate at the forward end of the pump. Stop the engine when the plate has moved out to its fullest extent, ie to the pump minimum stroke position. Repeat this if necessary and check that the pump pulley

3.1 The reed valve assembly is located here

has moved as far as possible clockwise: it is essential that the pump is fully in the minimum stroke position before any adjustments are made.

4 The pump minimum stroke setting is adjusted using shims, as described in Chapter 2; no variations have been made.

5 The oil pump cable adjustment should be checked only after the carburettors have been adjusted and synchronised, the throttle cable adjustment checked and the pump minimum stroke setting checked. The pump must be fully in its minimum stroke position as described above. The alignment of the pulley mark with the pump guide pin is checked with the throttle in one of two places. On early models, the throttle should be fully closed, but the twistgrip rotated just far enough to remove all traces of free play from the cables, while on later models the throttle must be fully open (Note: ignore the throttle slide marks, which are for carburettor synchronisation only).

6 Unfortunately it is not possible to be precise about when the modification was made. It would seem to have been introduced in 1976 and was certainly in general use by 1978, but note that this refers to the date of production, which is not necessarily the same as the date of registration. It should be obvious which is the correct position due to the relative position of the guide pin and pulley marks. Adjustment is made using the pump cable adjuster.

7 The problem is exacerbated by the fact that while the pulley mark should be rectangular with a raised triangular cross-section, as shown in Chapter 2, marks have been made in the form of circular drilled indentations, raised pips, or stamped lines. It is for this reason that it is so essential to obtain expert advice from a good Yamaha dealer if there is any doubt about the throttle position or pulley marks to be used on your machine.

6 CDI system

Ignition timing adjustment

1 Remove its pinch bolt and pull the gearchange lever off its shaft. Remove the crankcase left-hand cover. Remove the spark plug from the left-hand cylinder and screw in a dial gauge adaptor, then fit the gauge.

2 Using a spanner on the generator rotor nut, slowly rotate the engine anticlockwise whilst observing the gauge needle. Stop when TDC is reached. At this point zero the gauge and rock the crankshaft backwards and forwards slightly to check that the needle does not go past zero, resetting the gauge if necessary.

3 Rotate the crankshaft clockwise until a gauge reading of 3 – 4 mm is obtained, then slowly rotate it anticlockwise until it reads exactly 1.6 mm. This is the piston position before TDC at which the spark must occur. If the ignition timing is correct, the F1 mark stamped in the rotor periphery will align exactly with the fixed index mark stamped on the stator baseplate. The fixed index mark will be found in the 2 o'clock position, stamped either on a raised extension of the baseplate, or on the pulser coil, depending on the make of generator fitted. Note that a tolerance of 0.15 mm is allowed on each side of the piston position, so that if the marks align with the piston anywhere between 1.45 – 1.75 mm BTDC the timing is sufficiently accurate and need not be disturbed. If adjustment is required, hold the crankshaft in position, whilst the baseplate retaining screws are slackened slightly. Using a flat-bladed screwdriver move the baseplate to the required position by using the slots provided in the 7 o'clock position. Tighten the baseplate screws and recheck the timing as described above, making subsequent adjustment if necessary.

4 When the timing has been correctly set, refit the crankcase left-hand cover and gearchange lever. Remove the dial gauge and adaptor and refit the spark plug.

Generator coil testing

5 To check the condition of the pulser coil and ignition source

coil it will be necessary to use an ohmmeter or multimeter set to the resistance range. Trace the generator wiring up from the left-hand crankcase cover and disconnect it at the block connectors. Make the following tests on the generator side of the wiring.

Ignition source coil:
Red to brown wire .... 5.1 ohm ± 10% @ 20°C (68°F)
Brown to black wire .... 271 ohm ± 10% @ 20°C (68°F)
Pulser coil:
White/red to black wire ...... 87 ohm ± 10% @ 20°C (68°F)

6 If the test results are widely different from these figures, or if an open or short circuit is indicated it can be assumed that the coil is faulty and should be renewed. Check first, however, that this is not caused by a wire breakage or poor connection and always have your findings confirmed by a Yamaha dealer.

Ignition HT coil testing

7 Disconnect the ignition HT coil low tension leads at the block connector. Also disconnect the HT leads from the spark plugs and remove the suppressor caps from the leads. Make resistance tests across the primary and secondary windings using an ohmmeter or multimeter set to the resistance range. In the case of the primary windings connect the meter between the orange and black low tension wires on the coil side of the block connector; the specified resistance is 0.33 ohm ± 20% @ 20°C (68°F). Next measure the resistance across the two HT leads to check the secondary windings; the specified resistance is 3.5 K ohm ± 30% @ 20°C (68°F). If either reading is well outside the figures given a fault is indicated; have your findings confirmed by a Yamaha dealer, who will be able to test the coil on a spark gap tester.

CDI unit

8 No figure or test data is available with which to test the CDI unit. All that can be done if failure is suspected is to substitute the unit with one in known good condition. Before doing so, check the other system components as described above, not overlooking the associated wiring and switches.

Generator removal and refitting

9 If access to the generator coils is required the generator rotor must be removed. To do so safely, without damaging the coils, will require the use of a special extractor of the centre bolt type. This tool is available from Yamaha or a pattern version can be obtained from most motorcycle dealers.

10 Remove the gearchange lever and crankcase left-hand cover. Remove the generator centre nut, noting that it will be necessary to prevent engine rotation whilst the nut is slackened. If the cylinder head, barrels and pistons have been removed, pass a close-fitting round bar through the connecting rod eyes, resting the bar on blocks of wood placed across the crankcase mouth. If the engine is in the frame, lock the gear train by placing the machine in gear with the rear wheel touching the ground, and apply the rear brake. Remove the generator nut and screw in the extractor outer body. Tighten its centre bolt to draw the rotor off the crankshaft taper. If the rotor is reluctant to move, tap smartly on the extractor bolt head and try again.

11 If stator removal is required, mark its position in relation to the casing so that subsequent retiming will be easier, then remove its retaining screws. Disconnect its main wiring at the block connector and release the wire to the neutral switch.

12 When refitting, ensure that the rotor locates correctly with the Woodruff key in the crankshaft end and tighten the retaining nut securely. Check the ignition timing as described above.

7 Front fork assembly

On the 1978 RD250DX model the fork travel has been changed from 120 to 140 mm and the inner tube diameter from 34 mm to 35 mm.

8 Front and rear disc brakes: general description

1 There has been an increase in the performance of the reed valve induction models which has necessitated an improvement in braking efficiency and consequently the new RD250 and RD350 models have been fitted with an hydraulically operated front disc brake of the fixed caliper type. Later RD250DX models are fitted with hydraulically operated disc brakes on both the front and rear wheels. The floating caliper type of disc brake is used on the later models and requires a different overhaul procedure. For this reason, before attempting to remove any part of either the front or rear disc brake, check which type is fitted to the machine. Both types work on the same principle, the fixed type of caliper has two pistons and the floating type has only one. The right-hand side of the handlebar carries the master cylinder for the front disc brake and which forms an integral part of the front brake lever. The rear master cylinder is situated behind the right-hand side panel.

2 The front brake caliper unit is attached to the right-hand fork leg, and the disc to the right-hand side of the front wheel hub. The rear brake caliper unit is attached to the right-hand side of the swinging arm and the disc is also on the right-hand side. The master cylinders are interconnected with their respective caliper units by a brake hose and pipe, which is filled with the hydraulic fluid necessary to transmit the braking action from either the handlebar lever or the brake pedal to the brake caliper.

3 When the front brake lever is depressed, it causes the master piston to move within the master cylinder, to which it is linked. As the piston moves in the cylinder, it traps the brake fluid causing a pressure build-up which is transmitted to the caliper via the brake hose and pipe that forms the connecting link. The pressure in the caliper cylinders causes the brake pads of a fixed type caliper to move in their respective housings and bear against the disc, the friction between the two pads and the disc providing the braking action. As the brake lever is released, the pressure falls and the pistons return to their original positions. The pads no longer bear on the disc and the wheel is again free to revolve. If the caliper is of the floating type it has only one piston. When the single piston bears upon the disc, the caliper unit slides fractionally to the right, to bring the second, fixed, pad into use. This provides adequate pressure on the disc surfaces without the added complexity of a double piston caliper. The hydraulic pressure is supplied in the same way as the fixed caliper type. The leverage of the brake lever is such that it produces a force at the master cylinder piston approximately four times that applied to the brake lever itself. This is one of the reasons why the hydraulic braking system is thought to be more efficient than the old and more conventional drum brake.

9 Front disc brake (fixed caliper type): removing and replacing the disc and pads

1 The brake disc, attached to the right-hand side of the front wheel hub by eight bolts, rarely requires attention. Check for run out, which may have occurred as the result of crash damage, and for wear. Run out should not exceed 0.15 mm at any point and the disc itself must not be permitted to wear below the limit thickness of 6.5 mm. If these figures are exceeded in either case, the disc must be renewed.

2 The disc bolts to a disc bracket, which itself bolts to the right-hand side of the wheel hub. It is necessary to detach the front wheel from the machine by first placing the machine on the centre stand so that the front wheel is raised clear of the ground. Slacken the clamp bolt at the base of the lower, left-hand fork leg and withdraw the split pin from the front wheel spindle nut, in this case on the right-hand side of the machine. Disconnect the speedometer drive cable. Slacken and remove the spindle nut, then withdraw the spindle; the wheel can now be withdrawn from the forks. It is preferable to insert a wooden wedge between the brake pads at this stage, to prevent them from being expelled if the brake lever is inadvertently operated. Remove the eight bolts from the inner edge of the brake disc and remove the disc. Note that each bolt is fitted with a lock washer, which must not be omitted during reassembly.

3 The two brake pads have an 'ear' moulded on the side, which locates with a slot in the caliper. To remove the pads, insert a screwdriver into the slot and prise each pad out of position.

4 The pads are moulded from a special resin impregnated asbestos compound and if renewal is necessary, only the correct replacement should be fitted. Each pad has an overall thickness of 9.0 mm (0.354 inches) when new and has a red line painted around the periphery, which represents the wear limit. The wear limit is 4.5 mm (0.177 inches) and on no account should the pad be allowed to wear beyond this limit. Each pad is fitted with a protruding ear marked 'indicator' which allows the condition of the pads to be ascertained without any need to remove them. Using a feeler gauge check the distance between the indicator 'ear' inside face and the face of the disc. If the distance is less than 0.5 mm (0.01969 inches) on either pad, both pads must be renewed. It is preferable to remove the front wheel before the pads can be renewed although the caliper mounting bolts can be slackened so that the unit can be swung clear of the disc as an alternative. Do NOT apply the brake in an attempt to displace the pads. If the actuating pistons move beyond their normal limit of travel, air will be admitted to the hydraulic system, necessitating a complete bleed of the system when reassembly is completed.

5 Reassembly is accomplished by reversing the procedure used for dismantling. Make sure that the brake pads are correctly located in the caliper and that the shim fitted behind each pad is not omitted. Check that the front wheel revolves quite freely when reassembly is complete. Always check the brake action before taking the machine on the road.

10 Front disc brake (fixed caliper type): removing, renovating, and replacing the caliper unit

1 Before the caliper assembly can be removed from the right-hand fork leg, it is first necessary to drain off the hydraulic fluid. Disconnect the brake pipe at the union connection it makes with the caliper unit and allow the fluid to drain into a clean container. It is preferable to keep the front brake lever applied throughout this operation, to prevent the fluid from leaking out of the reservoir. A thick rubber band cut from a section of inner tube will suffice, if it is wrapped tightly around the lever and the handlebars.

2 Note that brake fluid is an extremely efficient paint stripper. Take care to keep it away from any paintwork on the machine or from any clear plastic, such as that sometimes used for instrument glasses.

3 When the fluid has drained off, remove the caliper mounting bolts and nuts, then rotate the caliper unit upwards and lift it

away from the disc and the machine. The brake pads can now be removed, using a screwdriver to prise them from their respective housings. Remove the two bolts that hold the two sections of the caliper unit together and when they have separated, the seal from the brake fluid inlet.

4 To displace the pistons, apply a blast of compressed air through the brake fluid inlet of each caliper section. Take care to catch each piston as it emerges from its bore - if dropped or prised out of position with a screwdriver, it may be damaged irreparably and will have to be replaced. Remove the piston seal and dust seal from each caliper section.

5 The parts removed should be cleaned thoroughly, using only brake fluid as the liquid. Petrol, oil or paraffin will cause the various seals to swell and degrade, and should not be used under any circumstances. When the various parts have been cleaned, they should be stored in polythene bags until reassembly, so that they are kept dust free.

6 Examine the pistons for score marks or other imperfections. If they have any imperfections they must be renewed, otherwise air or hydraulic fluid leakage will occur, which will impair braking efficiency. With regard to the various seals, it is advisable to renew them all, irrespective of their appearance. It is a small price to pay against the risk of a sudden and complete front brake failure. It is standard Yamaha practice to renew the seals every two years, even if no braking problems have occurred.

7 Reassemble under clinically-clean conditions, by reversing the dismantling procedure. Renew the caliper unit bridge bolts as a safety precaution, even if they appear undamaged. Reconnect the hydraulic fluid pipe and make sure the union has been tightened fully. Before the brake can be used, the whole system must be bled of air, by following the procedure described in Section 14 of this Chapter.

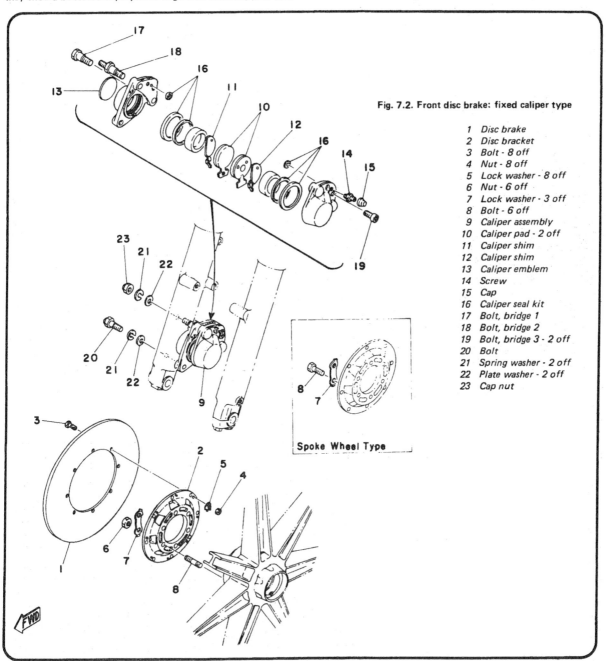

Fig. 7.2. Front disc brake: fixed caliper type

1 Disc brake
2 Disc bracket
3 Bolt - 8 off
4 Nut - 8 off
5 Lock washer - 8 off
6 Nut - 6 off
7 Lock washer - 3 off
8 Bolt - 6 off
9 Caliper assembly
10 Caliper pad - 2 off
11 Caliper shim
12 Caliper shim
13 Caliper emblem
14 Screw
15 Cap
16 Caliper seal kit
17 Bolt, bridge 1
18 Bolt, bridge 2
19 Bolt, bridge 3 - 2 off
20 Bolt
21 Spring washer - 2 off
22 Plate washer - 2 off
23 Cap nut

Spoke Wheel Type

11 Front disc brake (floating caliper type): removing and replacing the disc and pads

1 The brake disc. attached to the right-hand side of the front wheel hub by six bolts, rarely requires attention. Check for run out, which may have occurred as the result of crash damage, and for wear. Run out should not exceed 0.15 mm (0.006 in) at any point and the disc itself must not be permitted to wear below the limit thickness of 6.5 mm (0.255 in). If these figures are exceeded in either case, the disc must be renewed.

2 The disc bolts to a disc bracket, which itself bolts to the right-hand side of the wheel hub. It is necessary to detach the front wheel from the machine by first placing the machine on the centre stand so that the front wheel is raised clear of the ground. Slacken the clamp bolt at the base of the lower, left-hand fork leg and withdraw the split pin from the front wheel spindle nut, in this case on the right-hand side of the machine. Disconnect the speedometer drive cable. Slacken and remove the spindle nut, then withdraw the spindle; the wheel can now be withdrawn from the forks. It is preferable to insert a wooden wedge between the brake pads at this stage, to prevent them from being expelled if the brake lever is inadvertently operated. Remove the eight bolts from the inner edge of the brake disc and remove the disc. Note that each bolt is fitted with a lock washer, which must not be omitted during reassembly.

3 Pad removal may be accomplished without detaching the caliper unit from the fork leg if the front wheel has been removed. If wheel removal has not taken place the caliper unit must be removed. Disconnection of the hydraulic hose is not, however, required. The caliper may be detached by displacing the plastic cap and withdrawing the swing bolt upon which the caliper unit pivots.

4 The pads are secured by a pin which is located by a coil spring. Displace the spring so that the pin can be withdrawn, and lift out the pads. Examine each pad for wear. If it is found that either pad has worn down to the wear limit groove on the pad periphery or has an overall thickness of less than 6.5 mm (0.26 in) both pads should be renewed.

5 Reassembly may be carried out by reversing the dismantling procedure. The manufacturer recommends that the pad retaining pin and its locating spring is renewed when the pads are renewed as a safety precaution. If new pads are fitted some difficulty may be encountered when placing the assembled caliper unit over the disc, due to the reduction in the gap between the pad faces. If this proves to be the case the moving pad may be pushed inwards against the piston so that the piston moves back and the necessary clearance is obtained. The caliper swing bolt should be lubricated with grease before it is inserted and tightened.

6 The brakes **must** be checked for satisfactory operation before the machine is taken on the road.

12 Front disc brake (floating caliper type): removing, renovating, and replacing the caliper unit

1 Before the caliper assembly can be removed from the right-hand fork leg, it is first necessary to drain off the hydraulic fluid. Disconnect the brake pipe at the union connection it makes with the caliper unit and allow the fluid to drain into a clean container. It is preferable to keep the front brake lever applied throughout this operation, to prevent the fluid from leaking out of the reservoir. A thick rubber band cut from a section of inner tube will suffice, if it is wrapped tightly around the lever and the handlebars.

2 Note that brake fluid is an extremely efficient paint stripper. Take care to keep it away from any paintwork on the machine or from any clear plastic, such as that sometimes used for instrument glasses.

3 When the brake fluid has drained, remove the plastic cap and swing bolt to allow the caliper to be withdrawn from the machine. Remove the brake pads as described in the preceding Section.

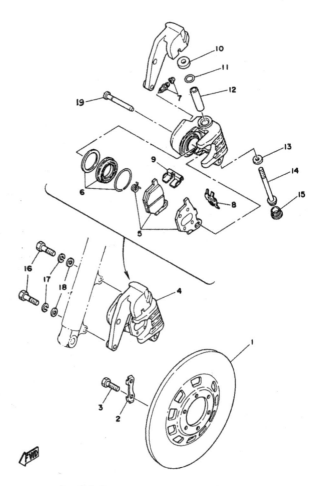

Fig. 7.3. Front disc brake: floating caliper type

1   Disc brake
2   Lock washer - 3 off
3   Bolt - 6 off
4   Front caliper assembly
5   Brake pad set
6   Caliper seal kit
7   Bleed screw and cap
8   Spring pad
9   Retainer
10 Plate washer
11 'O' ring set
12 Sleeve
13 Plate washer
14 Swing bolt
15 Swing bolt cap
16 Bolt - 2 off
17 Spring washer - 2 off
18 Plain washer - 2 off
19 Pad retaining pin

4 Withdraw the circlip and piston dust seal. To displace the piston, apply a blast of compressed air through the brake fluid inlet. Take care to catch the piston as it emerges from its bore — if dropped or prised out of position with a screwdriver, it may be damaged irreparably and will have to be replaced. Remove the piston seal from the caliper.

5 The parts removed should be cleaned thoroughly, using only brake fluid as the liquid. Petrol, oil or paraffin will cause the various seals to swell and degrade, and should not be used under any circumstances. When the various parts have been cleaned, they should be stored in polythene bags until re-assembly, so that they are kept dust free.

6 Examine the piston for score marks or other imperfections. If it has any imperfections it must be renewed, otherwise air or hydraulic fluid leakage will occur, which will impair braking efficiency. With regard to the various seals, it is advisable to renew them all, irrespective of their appearance. It is a small price to pay against the risk of a sudden and complete brake failure. It is standard Yamaha practice to renew the seals every two years, even if no braking problems have occurred.

7 Reassemble under clinically-clean conditions, by reversing the dismantling procedure. Reconnect the hydraulic fluid pipe and make sure the union has been tightened fully. Before the brake can be used, the whole system must be bled of air, by following the procedure described in Section 14 of this Chapter.

13 Master cylinder: examination and renewing seals

1 The master cylinder and hydraulic fluid reservoir take the form of a combined unit mounted on the right-hand side of the handlebars, to which the front brake lever is attached.

2 Before the master cylinder unit can be removed and dis-mantled, the system must be drained. Place a clean container below the brake caliper unit and attach a plastic tube from the bleed screw of the caliper unit to the container. Open the bleed screw one complete turn and drain the system by operating the brake lever until the master cylinder reservoir is empty. Close the bleed screw and remove the tube.

3 Before dismantling the master cylinder, it is essential that a clean working area is available on which the various component parts can be laid out. Use a sheet of white paper, so that none of the smaller parts can be overlooked.

4 Disconnect the stop lamp switch and front brake lever, taking care not to misplace the brake lever return spring. The stop lamp switch (if fitted) is attached to the bolt that acts as the brake lever pivot. Remove the split pin and castellated nut and take off the switch. Remove the brake hose by unscrewing the banjo union bolt. Take the master cylinder away from the handlebars by removing the two bolts that clamp it to the handlebars. Check that all fluid has drained from the reservoir, by taking off the reservoir cap and the diaphragm below.

5 Withdraw the rubber boot that protects the end of the master cylinder and remove the snap ring that holds the piston assembly in position, using a pair of circlip pliers. The piston assembly can now be drawn out, and the spring below it.

6 To dismantle the piston assembly, remove the 'E' clip at the far end and then the cylinder cup retainer. The cylinder cup can then be detached.

7 Examine the piston and the cylinder cup very carefully. If either is scratched or has the working surface impaired in any other way, it must be renewed without question. Reject the various seals, irrespective of their condition, and fit new ones in their place. It often helps to soften them a little before they are fitted by immersing them in a container of clean brake fluid.

8 When reassembling, follow the dismantling procedure in reverse, but take great care that none of the component parts is scratched or damaged in any way. Use brake fluid as the lubricant whilst reassembling. When assembly is complete, re-connect the brake fluid pipe and tighten the banjo union bolt to the recommended setting. Add about 30cc of brake fluid to the reservoir and bleed the system of air by following the procedure described in Section 14 of this Chapter.

13.1 Master cylinder is attached to the front brake lever

14 Bleeding the hydraulic system

1 As mentioned earlier, brake action is impaired or even rendered inoperative if air is introduced into the hydraulic system. This can occur if the seals leak, the reservoir is allowed to run dry or if the system is drained prior to the dismantling of any component part of the system. Even when the system is refilled with hydraulic fluid, air pockets will remain and because air will compress, the hydraulic action is lost.

2 Check the fluid content of the reservoir and fill almost to the top. Remember that hydraulic brake fluid is an excellent paint stripper, so beware of spillage, especially near the petrol tank.

3 Place a clean glass jar below the brake caliper unit and attach a clear plastic tube from the caliper bleed screw to the container. Place some clean hydraulic fluid in the container so that the pipe is always immersed below the surface of the fluid.

4 Unscrew the bleed screw one complete turn and pump the handlebar lever slowly. As the fluid is ejected from the bleed screw the level in the reservoir will fall. Take care that the level does not drop too low whilst the operation continues, otherwise air will re-enter the system, necessitating a fresh start.

5 Continue the pumping action with the lever until no further air bubbles emerge from the end of the plastic pipe. Hold the brake lever against the handlebars and tighten the caliper bleed screw. Remove the plastic tube AFTER the bleed screw is closed.

6 Check the brake action for sponginess, which usually denotes there is still air in the system. If the action is spongy, continue the bleeding operation in the same manner, until all traces of air are removed.

7 Bring the reservoir up to the correct level of fluid and replace the diaphragm, sealing gasket and cap. Check the entire system for leaks. Recheck the brake action.

8 Brake fluid drained from the system will almost certainly be contaminated, either by foreign matter or more commonly by the absorption of water from the air. All hydraulic fluids are to some degree hygroscopic, that is, they are capable of drawing water from the atmosphere, and thereby degrading their specifi-cations. In view of this, and the relative cheapness of the fluid, old fluid should always be discarded.

15 Hydraulic brake hose and pipe: examination

1   An external brake hose and pipe is used to transmit the hydraulic pressure to the caliper unit when the front brake or rear brake is applied. The brake hose is of the flexible type, fitted with an armoured surround. It is capable of withstanding pressures up to 350 kg/cm^2. The brake pipe attached to it is made from double steel tubing, zinc plated to give better corrosion resistance.
2   When the brake assembly is being overhauled, check the condition of both the hose and the pipe for signs of leakage or scuffing, if either has made rubbing contact with the machine whilst it is in motion. The union connections at either end must also be in good condition, with no stripped threads or damaged sealing washers. Check also the feed pipe from the rear brake master cylinder to the reservoir. This pipe is not subjected to pressure but may perish after a considerable length of time. It is a push fit on the unions and is retained by screw clips.

16 Front wheel: examination and renovation (cast alloy wheel models)

1   Carefully check the complete wheel for cracks and chipping, particularly at the spoke roots and the edge of the rim. As a general rule a damaged wheel must be renewed as cracks will cause stress points which may lead to sudden failure under heavy load. Small nicks may be radiused carefully with a fine file and emery paper (No. 600 - No. 1000) to relieve the stress. If there is any doubt as to the condition of a wheel, advice should be sought from a Yamaha repair specialist.
2   Each wheel is covered with a coating of lacquer, to prevent corrosion. If damage occurs to the wheel and the lacquer finish is penetrated, the bared aluminium alloy will soon start to corrode. A whitish grey oxide will form over the damaged area, which in itself is a protective coating. This deposit however, should be removed carefully as soon as possible and a new protective coating of lacquer applied.
3   Check the lateral run out at the rim by spinning the wheel and placing a fixed pointer close to the rim edge. If the maximum run out is greater than 1.0 mm (0.03937 in), Yamaha recommend that the wheel be renewed. This is, however, a council of perfection; a run out somewhat greater than this can probably be accommodated without noticeable effect on steering. No means is available for straightening a warped wheel without resorting to the expense of having the wheel skimmed on all faces. If warpage was caused by impact during an accident, the safest measure is to renew the wheel complete. Worn wheel bearings may cause rim run out. These should be renewed as described in Section 16 of this Chapter.

17 Wheel bearings: examination and replacement

1   Access to the front wheel bearings may be made after removal of the wheel from the forks. Pull the speedometer gearbox out of the hub left-hand boss and remove the dust seal cover and wheel spacer from the hub right-hand side.
2   Lay the wheel on the ground with the disc side facing downward and with a special tool, in the form of a rod with a curved end, insert the curved end into the hole in the centre of the spacer separating the two wheel bearings. If the other end of the special tool is hit with a hammer, the right-hand bearing, bearing flange washer, and bearing spacer will be expelled from the hub.
3   Invert the wheel and drive out the left-hand bearing by inserting a drift of the appropriate size, through the hub. During the removal of either bearing it may be necessary to support the wheel across an open-ended box so that there is sufficient

clearance for the bearing to be displaced completely from the hub.
4   Remove all the old grease from the hub and bearings, giving the latter a final wash in petrol. Check the bearings for signs of play or roughness when they are turned. If there is any doubt about the condition of a bearing, it should be renewed.
5   Before replacing the bearings, first pack the hub with new grease. Then drive the bearings back into position, not forgetting the distance piece that separates them. Take great care to ensure that the bearings enter the housings perfectly squarely otherwise the housing surface may be broached. Fit replacement oil seals and any dust covers or spacers that were also displaced during the original dismantling operation.

18 Front wheel: reassembly and replacement

1   Refit the speedometer gearbox to the left-hand side of the hub, ensuring that the drive dogs engage correctly.
2   Lift the wheel upwards so that the brake disc enters the brake caliper squarely and align the wheel so that the front wheel spindle can be inserted from the left. Push the spindle home through the right-hand fork leg until the head of the spindle is flush with the left-hand fork leg. Ensure that the axial slot on the speedometer gearbox is located with the projecting lug on the inside of the left-hand fork lower leg. This is most important otherwise the speedometer gearbox may revolve with the wheel and snap the drive cable. Replace the washer and castellated nut on the end of the fork spindle and tighten the nut. Replace the split pin that passes through the end of the nut and the spindle. Re-tighten the split clamp at the extreme end of the right-hand fork leg, by tightening the forward nut first, followed by the rear nut. A gap will remain between the rear face of the clamp and the bottom rear face of the fork leg.
3   Spin the wheel to ensure that it moves freely, then attach the speedometer drive cable, which is retained by a knurled ring. Make sure that the cable passes through the cable retaining loop of the mudguard, to ensure it cannot come into contact with either the tyre or wheel.

19 Rear disc brake: removing and replacing the disc and pads

1   The rear brake disc and disc bracket are identical to the components fitted to the front wheel. Removal and examination of the disc and also the brake pads may be carried out by referring to the information given in Section 10 of this Chapter.

20 Rear disc brake: removing, renovating and replacing the caliper unit

The caliper unit fitted to the rear brake is similar in all respects to that utilised on the front brake system. Removal and inspection is materially the same and should be carried out by following the procedure given in Section 11 of this Chapter.

21 Rear wheel: examination and bearing replacement

1   When inspecting the rear wheel follow the procedure given for front wheel inspection in either Chapter 5, Section 2 or Section 15 of this Chapter, depending upon whether the wheel is of the spoked or cast alloy type.
2   The procedure for removal, examination and lubrication of the rear wheel bearings is materially the same as that used when attending to the front wheel bearings. Follow the procedure given in Section 16 of this Chapter.

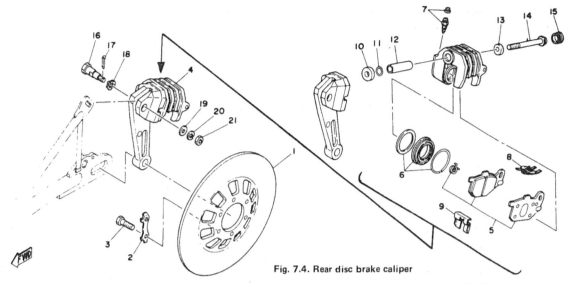

Fig. 7.4. Rear disc brake caliper

1 Disc brake	7 Bleed screw with cap	12 Sleeve	17 Split pin
2 Lock washer - 3 off	8 Spring pad	13 Plate washer	18 Washer
3 Bolt - 6 off	9 Retainer	14 Swing bolt	19 Plate washer
4 Rear caliper assembly	10 Plate washer	15 Swing bolt cap	20 Washer spring
5 Brake pad set	11 'O' ring	16 Bolt	21 Nut
6 Caliper seal kit			

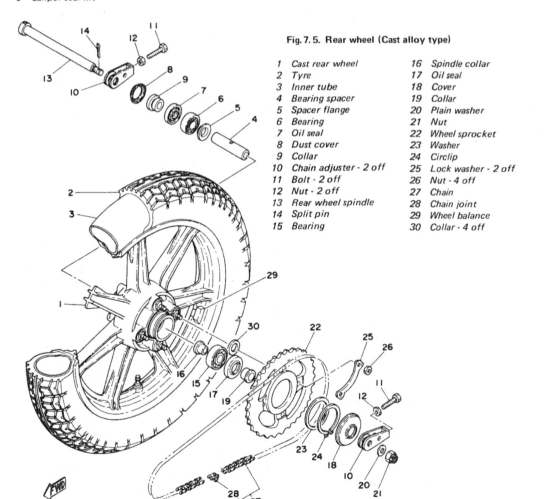

Fig. 7.5. Rear wheel (Cast alloy type)

1 Cast rear wheel	16 Spindle collar
2 Tyre	17 Oil seal
3 Inner tube	18 Cover
4 Bearing spacer	19 Collar
5 Spacer flange	20 Plain washer
6 Bearing	21 Nut
7 Oil seal	22 Wheel sprocket
8 Dust cover	23 Washer
9 Collar	24 Circlip
10 Chain adjuster - 2 off	25 Lock washer - 2 off
11 Bolt - 2 off	26 Nut - 4 off
12 Nut - 2 off	27 Chain
13 Rear wheel spindle	28 Chain joint
14 Split pin	29 Wheel balance
15 Bearing	30 Collar - 4 off

22 Rear brake master cylinder: removal, examination and renewing seals

1   The rear brake master cylinder is attached to the right-hand rear frame downtube and is operated from a foot pedal via a push rod connected to the pedal by a clevis fork and pin. The master cylinder reservoir is a separate component, remote from the master cylinder and interconnected by a short feed hose.

2   Drain the master cylinder and reservoir, using a similar technique to that described for the front brake master cylinder. The master cylinder reservoir is fitted with a screw cap.

3   Loosen the lower screw clip on the hydraulic fluid feed pipe and pull the pipe off the union. Take care not to drop any remaining fluid on the paintwork. Unscrew the reservoir retaining bolt and then lift the reservoir upwards off the support

bracket.

4   Disconnect the hydraulic hose at the brake caliper by removing the banjo bolt. The master cylinder is retained on the frame lug by two bolts. After removal of the bolt, the cylinder unit may be lifted upwards so that the operating push rod leaves the cylinder. Disconnect the brake lamp leads which are a push fit on the switch terminals. The master cylinder can now be lifted away from the machine.

5   Examination and dismantling of the rear brake master cylinder may be made by referring to the directions in Section 12 and 21 of this Chapter. Additionally, the reservoir should be flushed out with clean fluid before refitting.

6   After reassembly and replacement of the rear brake master cylinder components which may be made by reversing the dismantling procedure - bleed the rear brake system of air by referring to Section 13 of this Chapter.

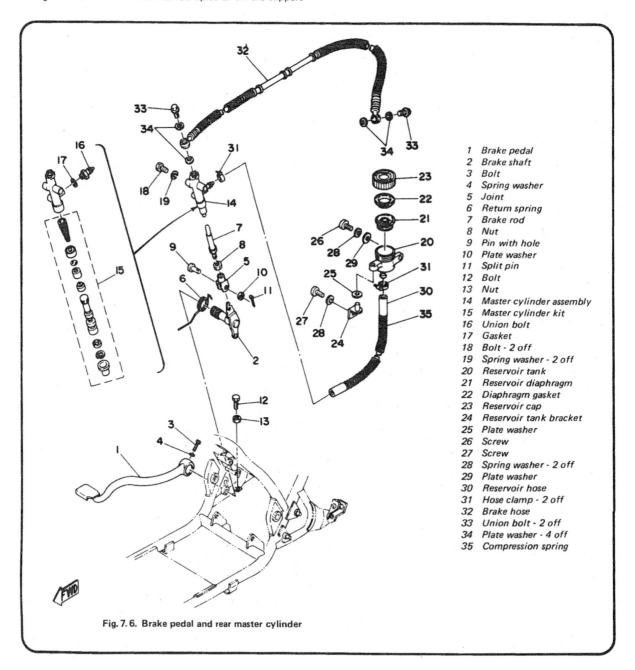

1   Brake pedal
2   Brake shaft
3   Bolt
4   Spring washer
5   Joint
6   Return spring
7   Brake rod
8   Nut
9   Pin with hole
10 Plate washer
11 Split pin
12 Bolt
13 Nut
14 Master cylinder assembly
15 Master cylinder kit
16 Union bolt
17 Gasket
18 Bolt - 2 off
19 Spring washer - 2 off
20 Reservoir tank
21 Reservoir diaphragm
22 Diaphragm gasket
23 Reservoir cap
24 Reservoir tank bracket
25 Plate washer
26 Screw
27 Screw
28 Spring washer - 2 off
29 Plate washer
30 Reservoir hose
31 Hose clamp - 2 off
32 Brake hose
33 Union bolt - 2 off
34 Plate washer - 4 off
35 Compression spring

Fig. 7.6. Brake pedal and rear master cylinder

22.1A Rear brake master cylinder retained by two bolts

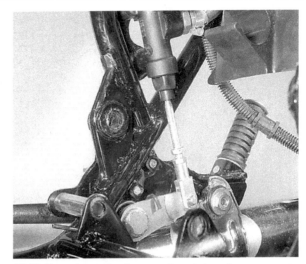

22.1B Operating pushrod passes through boot into rear master cylinder

23 Electrical equipment: indicator lamp display

1 Indicator lamp display.
   To provide better display facilities, the indicator lamps fitted to the RD250 and RD350 models, have been transferred to a display panel mounted between the tachometer and the speedometer heads and are no longer accommodated in the former. Visual indication is given when the stop lamp is operating, when the headlamp is on main beam, and which set of flashing indicators are in use.
2 The later RD250DX model has an indicator switch which is operated in the usual way by pushing it to either the left or the right, but it will either cancel itself after 15 seconds or it can be cancelled manually by depressing it with the thumb.
3 The lamp holders are a push fit into the base of the display panel, using a similar type of rubber encased mounting to that which was previously inserted into the base of the tachometer head.

24 Other minor modifications

   All other modifications are of a minor nature, improving the overall characteristics of the machines, though these alterations do not affect the technical content of this manual. They can be summarised as follows:
1 Rubber blocks are interposed between the cylinder fins to reduce oscillation noise.
2 Modified silencers have been introduced to reduce noise level without impairing performance.
3 Raised screwtops are fitted to the Mikuni carburettors.
4 A certain amount of redesign has been applied to the telescopic front forks.
5 A different type of sparking plug cap is used (1978 model).
6 The voltage regulator and rectifier are now both combined in the same unit (ie changed from the separate to the integrated type with effect from the 1978 model).

23.1 Display panel contains the main indicator lamps

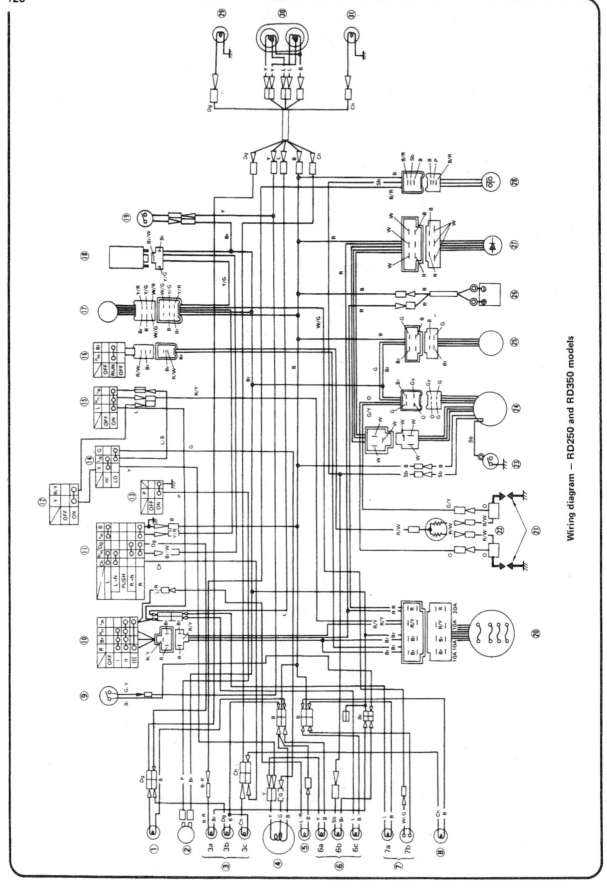

Wiring diagram — RD250 and RD350 models

KEY TO WIRING DIAGRAM

1	Right front indicator	13	Horn button
2	Horn	14	Dipswitch
3	Warning lamp cluster	15	Light switch
3a	Oil warning lamp	16	Engine cut switch
3b	Flasher warning lamp (right)	17	Cancelling unit
3c	Flasher warning lamp (left)	18	Flasher relay
4	Headlamp	19	Rear brake stop lamp switch
5	Parking lamp	20	Fuse box
6	Tachometer	21	Sparking plug
6a	Main beam warning lamp	22	Ignition coil
6b	Neutral warning lamp	23	Neutral indicator switch
6c	Tachometer illumination lamp - 2 off	24	ac generator
7	Speedometer	25	Regulator
7a	Speedometer illumination lamp - 2 off	26	Battery
7b	Odometer sender	27	Rectifier
8	Left front indicator	28	Oil level switch
9	Front brake stop switch	29	Rear indicator lamp (right)
10	Main switch	30	Tail/stop lamp
11	Direction indicator switch	31	Rear indicator lamp (left)
12	Headlamp flasher switch		

Colour code

R	Red	P	Pink
O	Orange	Y	Yellow
G	Green	Dg	Dark green
Br	Brown	Ch	Dark brown (chocolate)
B	Black	L	Blue
Sb	Sky blue	W	White
Gy	Grey	R/Y	Red/Yellow
R/W	Red/White	Y/R	Yellow/Red
W/R	White/Red	W/G	White/Green
G/Y	Green/Yellow	L/B	Blue/Black
L/R	Blue/Red	Br/W	Brown/White
B/R	Black/Red		

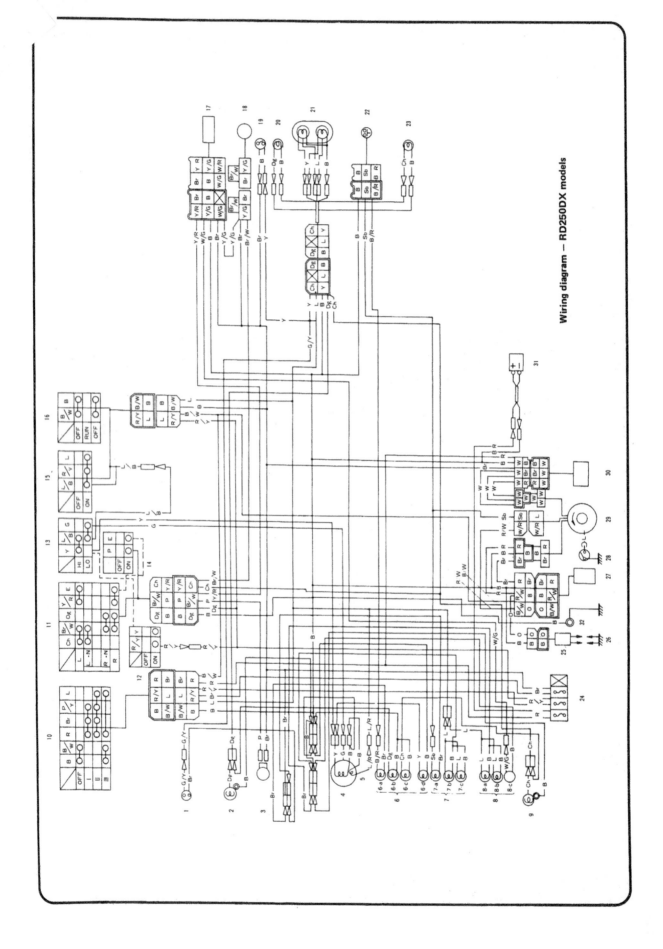

Wiring diagram — RD250DX models

KEY TO WIRING DIAGRAM

1   Front brake stop switch
2   Front indicator flasher (right)
3   Horn
4   Headlamp
5   Parking lamp
6   Warning lamp cluster
6a   Oil warning lamp
6b   Flasher warning lamp (right)
6c   Flasher warning lamp (left)
6d   Neutral warning lamp
7   Tachometer
7a   Main beam
7b   Tachometer illumination lamp
7c   Tachometer illumination lamp
8   Speedometer
8a   Speedometer illumination lamp
8b   Speedometer illumination lamp
8c   Odometer sender
9   Front indicator flasher (left)
10   Main switch
11   Direction indicator switch

12   Headlamp flasher
13   Dipswitch
14   Horn button
15   Light switch
16   Engine cut switch
17   Cancelling unit
18   Flasher relay
19   Rear brake stop switch
20   Rear indicator flasher (right)
21   Tail/brake light
22   Oil level switch
23   Rear indicator flasher (left)
24   Fuse box
25   Ignition coil
26   Sparking plug
27   C.D.I. unit
28   Neutral indicator switch
29   ac Magneto
30   Regulator and rectifier (integrated type)
31   Battery
32   Earth

Colour code

R   Red
O   Orange
G   Green
Br   Brown
B   Black
Sb   Sky blue
Gy   Grey
R/W Red/White
W/R White/Red
G/Y Green/Yellow
L/R Blue/Red
B/R Black/Red

P   Pink
Y   Yellow
Dg   Dark green
Ch   Dark brown (chocolate)
L   Blue
W   White
R/Y Red/Yellow
Y/R Yellow/Red
Y/G Yellow/Green
W/G White/Green
L/B Blue/Black
Br/W Brown/White

Metric conversion tables

Inches	Decimals	Millimetres
1/64	0.015625	0.3969
1/32	0.03125	0.7937
3/64	0.046875	1.1906
1/16	0.0625	1.5875
5/64	0.078125	1.9844
3/32	0.09375	2.3812
7/64	0.109375	2.7781
1/8	0.125	3.1750
9/64	0.140625	3.5719
5/32	0.15625	3.9687
11/64	0.171875	4.3656
3/16	0.1875	4.7625
13/64	0.203125	5.1594
7/32	0.21875	5.5562
15/64	0.234375	5.9531
1/4	0.25	6.3500
17/64	0.265625	6.7469
9/32	0.28125	7.1437
19/64	0.296875	7.5406
5/16	0.3125	7.9375
21/64	0.328125	8.3344
11/32	0.34375	8.7312
23/64	0.359375	9.1281
3/8	0.375	9.5250
25/64	0.390625	9.9219
13/32	0.40625	10.3187
27/64	0.421875	10.7156
7/16	0.4375	11.1125
29/64	0.453125	11.5094
15/32	0.46875	11.9062
31/64	0.48375	12.3031
1/2	0.5	12.7000
33/64	0.515625	13.0969
17/32	0.53125	13.4937
35/64	0.546875	13.8906
9/16	0.5625	14.2875
37/64	0.578125	14.6844
19/32	0.59375	15.0812
39/64	0.609375	15.4781
5/8	0.625	15.8750
41/64	0.640625	16.2719
21/32	0.65625	16.6687
43/64	0.671875	17.0656
11/16	0.6875	17.4625
45/64	0.703125	17.8594
23/32	0.71875	18.2562
47/64	0.734375	18.6531
3/4	0.75	19.0500
49/64	0.765625	19.4469
25/32	0.78125	19.8437
51/64	0.796875	20.2406
13/16	0.8125	20.6375
53/64	0.828125	21.0344
27/32	0.84375	21.4312
55/64	0.859375	21.8281
7/8	0.875	22.2250
57/64	0.890625	22.6219
29/32	0.90625	23.0187
59/64	0.921875	23.4156
15/16	0.9375	23.8125
61/64	0.953125	24.2094
31/32	0.96875	24.6062
63/64	0.984375	25.0031

Millimetres to Inches

mm	Inches
0.01	0.00039
0.02	0.00079
0.03	0.00118
0.04	0.00157
0.05	0.00197
0.06	0.00236
0.07	0.00276
0.08	0.00315
0.09	0.00354
0.1	0.00394
0.2	0.00787
0.3	0.01181
0.4	0.01575
0.5	0.01969
0.6	0.02362
0.7	0.02756
0.8	0.03150
0.9	0.03543
1	0.03937
2	0.07874
3	0.11811
4	0.15748
5	0.19685
6	0.23622
7	0.27559
8	0.31496
9	0.35433
10	0.39370
11	0.43307
12	0.47244
13	0.51181
14	0.55118
15	0.59055
16	0.62992
17	0.66929
18	0.70866
19	0.74803
20	0.78740
21	0.82677
22	0.86614
23	0.90551
24	0.94488
25	0.98425
26	1.02362
27	1.06299
28.	1.10236
29	1.14173
30	1.18110
31	1.22047
32	1.25984
33	1.29921
34	1.33858
35	1.37795
36	1.41732
37	1.4567
38	1.4961
39	1.5354
40	1.5748
41	1.6142
42	1.6535
43	1.6929
44	1.7323
45	1.7717

Inches to Millimetres

Inches	mm
0.001	0.0254
0.002	0.0508
0.003	0.0762
0.004	0.1016
0.005	0.1270
0.006	0.1524
0.007	0.1778
0.008	0.2032
0.009	0.2286
0.01	0.254
0.02	0.508
0.03	0.762
0.04	1.016
0.05	1.270
0.06	1.524
0.07	1.778
0.08	2.032
0.09	2.286
0.1	2.54
0.2	5.08
0.3	7.62
0.4	10.16
0.5	12.70
0.6	15.24
0.7	17.78
0.8	20.32
0.9	22.86
1	25.4
2	50.8
3	76.2
4	101.6
5	127.0
6	152.4
7	177.8
8	203.2
9	228.6
10	254.0
11	279.4
12	304.8
13	330.2
14	355.6
15	381.0
16	406.4
17	431.8
18	457.2
19	482.6
20	508.0
21	533.4
22	558.8
23	584.2
24	609.6
25	635.0
26	660.4
27	685.8
28	711.2
29	736.6
30	762.0
31	787.4
32	812.8
33	838.2
34	863.6
35	889.0
36	914.4

1   Imperial gallon = 8 Imp pints = 1.20 US gallons = 277.42 cu in = 4.54 litres

1   US gallon = 4 US quarts = 0.83 Imp gallon = 231 cu in = 3.78 litres

1   Litre = 0.21 Imp gallon = 0.26 US gallon = 61.02 cu in = 1000 cc

Miles to Kilometres		Kilometres to Miles	
1	1.61	1	0.62
2	3.22	2	1.24
3	4.83	3	1.86
4	6.44	4	2.49
5	8.05	5	3.11
6	9.66	6	3.73
7	11.27	7	4.35
8	12.88	8	4.97
9	14.48	9	5.59
10	16.09	10	6.21
20	32.19	20	12.43
30	48.28	30	18.64
40	64.37	40	24.85
50	80.47	50	31.07
60	96.56	60	37.28
70	112.65	70	43.50
80	128.75	80	49.71
90	144.84	90	55.92
100	160.93	100	62.14

lbf ft to kgf m		kgf m to lbf ft		lbf/in^2 to kgf/cm^2		kgf/cm^2 to lbf/in^2	
1	0.138	1	7.233	1	0.07	1	14.22
2	0.276	2	14.466	2	0.14	2	28.50
3	0.414	3	21.699	3	0.21	3	42.67
4	0.553	4	28.932	4	0.28	4	56.89
5	0.691	5	36.165	5	0.35	5	71.12
6	0.829	6	43.398	6	0.42	6	85.34
7	0.967	7	50.631	7	0.49	7	99.56
8	1.106	8	57.864	8	0.56	8	113.79
9	1.244	9	65.097	9	0.63	9	128.00
10	1.382	10	72.330	10	0.70	10	142.23
20	2.765	20	144.660	20	1.41	20	284.47
30	4.147	30	216.990	30	2.11	30	426.70

English/American terminology

Because this book has been written in England, British English component names, phrases and spellings have been used throughout. American English usage is quite often different and whereas normally no confusion should occur, a list of equivalent terminology is given below.

English	American	English	American
Air filter	Air cleaner	Number plate	License plate
Alignment (headlamp)	Aim	Output or layshaft	Countershaft
Allen screw/key	Socket screw/wrench	Panniers	Side cases
Anticlockwise	Counterclockwise	Paraffin	Kerosene
Bottom/top gear	Low/high gear	Petrol	Gasoline
Bottom/top yoke	Bottom/top triple clamp	Petrol/fuel tank	Gas tank
Bush	Bushing	Pinking	Pinging
Carburettor	Carburetor	Rear suspension unit	Rear shock absorber
Catch	Latch	Rocker cover	Valve cover
Circlip	Snap ring	Selector	Shifter
Clutch drum	Clutch housing	Self-locking pliers	Vise-grips
Dip switch	Dimmer switch	Side or parking lamp	Parking or auxiliary light
Disulphide	Disulfide	Side or prop stand	Kick stand
Dynamo	DC generator	Silencer	Muffler
Earth	Ground	Spanner	Wrench
End float	End play	Split pin	Cotter pin
Engineer's blue	Machinist's dye	Stanchion	Tube
Exhaust pipe	Header	Sulphuric	Sulfuric
Fault diagnosis	Trouble shooting	Sump	Oil pan
Float chamber	Float bowl	Swinging arm	Swingarm
Footrest	Footpeg	Tab washer	Lock washer
Fuel/petrol tap	Petcock	Top box	Trunk
Gaiter	Boot	Torch	Flashlight
Gearbox	Transmission	Two/four stroke	Two/four cycle
Gearchange	Shift	Tyre	Tire
Gudgeon pin	Wrist/piston pin	Valve collar	Valve retainer
Indicator	Turn signal	Valve collets	Valve cotters
Inlet	Intake	Vice	Vise
Input shaft or mainshaft	Mainshaft	Wheel spindle	Axle
Kickstart	Kickstarter	White spirit	Stoddard solvent
Lower leg	Slider	Windscreen	Windshield
Mudguard	Fender		

Index